THIRD EDITION

Hazardous Materials Monitoring and Detection Devices

Christopher Hawley

JONES & BARTLETT
LEARNING

Jones & Bartlett Learning
World Headquarters
5 Wall Street
Burlington, MA 01803
978-443-5000
info@jblearning.com
www.jblearning.com
www.psglearning.com

Jones & Bartlett Learning books and products are available through most bookstores and online booksellers. To contact the Jones & Bartlett Learning Public Safety Group directly, call 800-832-0034, fax 978-443-8000, or visit our website, www.psglearning.com.

18609-3

Production Credits
General Manager and Executive Publisher: Kimberly Brophy
VP, Product Development: Christine Emerton
Executive Editor: Bill Larkin
Vendor Manager: Molly Hogue
Associate Marketing Manager: Hayley Lorge
VP, Manufacturing and Inventory Control: Therese Connell
Composition: S4Carlisle Publishing Services
Project Management: S4Carlisle Publishing Services

Cover Design: Kristin E. Parker
Rights & Media Specialist: Thais Miller
Media Development Editor: Troy Liston
Cover Image (Title Page, Part Opener, Chapter Opener): © Jones & Bartlett Learning. Photographed by Glen E. Ellman.
Printing and Binding: LSC Communications
Cover Printing: LSC Communications

Library of Congress Cataloging-in-Publication Data

Names: Hawley, Chris, author.
Title: Hazardous materials monitoring and detection devices / Christopher Hawley.
Other titles: Hazardous materials air monitoring and detection devices
Description: Third edition. | Burlington, MA : Jones & Bartlett Learning, [2020] | Previous edition bears title: Hazardous materials air monitoring and detection devices. | Includes bibliographical references and index.
Identifiers: LCCN 2018022624 | ISBN 9781284143911 (perfect : alk. paper)
Subjects: LCSH: Hazardous substances. | Air--Pollution--Measurement. | Air sampling apparatus. | Chemical detectors.
Classification: LCC T55.3.H3 H377 2019 | DDC 628.5/30287--dc23 LC record available at https://lccn.loc.gov/2018022624

6048

Printed in the United States of America
22 21 20 19 18   10 9 8 7 6 5 4 3 2 1

Dedication

This text is dedicated to
all the emergency responders
who respond every day and
put it on the line
for their community.

Brief Contents

Preface

Hazardous Materials: Monitoring and Detection Devices, Third Edition, is designed for a variety of industries. Although primarily written for emergency responders, hazardous materials responders, fire fighters, and law enforcement officers, the text applies to several other occupations. Persons who work in an industrial facility or who are involved in health and safety, such as industrial hygienists or safety managers, will find this text very helpful. Persons involved in environmental recovery or in other areas where monitoring is used will also benefit. This text covers monitors and detection devices for both hazardous materials and weapons of mass destruction (WMD). This text provides these agencies with a broad-spectrum picture of monitoring, one that can help with purchasing decisions and in the implementation of a monitoring strategy. This text covers a wide variety of detection devices, some basic and some advanced.

An important part of this book it is how to use these devices tactically and how to interpret the readings. The backbone of the text is the discussion of risk-based response (RBR), which is a common approach to emergency response. Many response agencies follow a risk-based response, and the NFPA 472 standard includes the recommendation to follow this method. Since this standard added risk-based response, several risk methodologies have been developed and are in common use. Every response team should adopt one or more methodologies for effective response. A lot of work went into developing these response methodologies. Using any of these methodologies can be a dramatic change and sometimes requires thinking outside the conventional hazardous materials box. For dealing with rescues, public endangerment, and terrorism, responders should use a safe, flexible, and efficient response system.

Why I Wrote This Book

As an instructor, I struggle with trying to develop a methodology to provide my knowledge to the reader. Trying to keep up with all the new detection devices is very difficult, even for someone involved in the business. To keep things simple, I focus on the technologies and how they function. If you understand how the technology works, you can compare apples to apples when discussing detection devices. Over the years, I have always thought of detection devices as tools in a tool box, and I always wanted to have a variety of tools. Knowing how they work and, more importantly, when they do not work is vital for your safety. Further, using a risk-based philosophy is vital for your safety and for that of the citizens you protect. *Hazardous Materials: Monitoring and Detection Devices, Third Edition*, covers the thought process behind RBR. The goal of RBR is to assist the responder in making

appropriate decisions regarding response tactics. RBR originated from several sources, and I am not the sole originator. Initially, Frank Docimo laid the groundwork with street smart concepts that simplified hazardous materials response. Mike Callan's thought process of "safe, unsafe, and dangerous" added to this process by simplifying the potential risk in a situation. Later these concepts were expanded by me and Buzz Melton, who at the time was the hazardous materials chief for the Baltimore City Fire Department and later an environmental chemist with FMC Baltimore. We developed technician and technician refresher programs for both of our departments, which included the risk-based response idea. We tried many ideas and eventually settled on a concept that is widely used today. Many hazardous materials and WMD response teams use an RBR profile on a regular basis, and it is part of their daily response profile.

Acknowledgments

There is no possible way to thank all those who have helped me throughout my career. I could provide a listing of names but would invariably miss some people. There are some persons who helped me with this text, who cannot be named, and I appreciate their input. With that said, the first people I wish to thank is my family, especially my wife, Donna, and my sons, Chris, Timothy, and Matthew, who are a constant source of inspiration and fun. Donna bears the brunt of putting up with my deadlines and scrambling. She is to be thanked for taking care of a lot of things while I do the traveling road show. Several friends have been key in helping me with my road show, career, and authoring this and my other texts: John Eversole (RIP), Steve Patrick, Greg Noll, Mike Hildebrand, Mike Callan, Richard Brooks, Luther Smith, Gary Warren, Chris Wrenn, and my road partner Paul (PJ) Cusic. All the personnel assigned to the FBI THRU, the hazardous materials officers, scientists, and special agents deserve thanks; they have been an inspiration and very helpful in the writing of this text. They have always been there when I needed a quick review of a thought, a topic, or a wild idea. Geoff Donahue and his team at Maryland Department of the Environment (MDE) are always a phone call away. They are by far one the best hazardous materials teams I have run across, and one of their team members, Robert Swann, is one of the most knowledgeable responders; he was an enormous help in the development of all my texts and always has answers to my questions. The other responders from MDE deserve my thanks as well for all their assistance. Armando (Toby) Bevelaqua got me through some serious writer's block; the flow charts in Chapter 12 were finished through his assistance and advice. The personnel assigned to the hazardous materials team at Station 14 deserve a lot of credit for their hard work and patience. As always, make sure you have fun (rule #1), don't get killed (rule #2), always ask questions, and never stop learning (rule #3). Please feel free to contact me directly with comments, suggestions, or questions: chawleym@gmail.com.

Always be safe!

Chris Hawley

AUTHOR NOTE: When using monitoring and detection devices for real-world situations, protective clothing is always recommended. In some of the photographs used in this text, you may see a person holding a device without gloves or other protective clothing. These photographs were not taken during an emergency response but only to show the device. No individuals were harmed during the development of this text.

The author and Jones & Bartlett Learning would like to thank the following persons for their review and assistance with this text:

Glen Rudner, Hazardous Materials Officer, Norfolk Southern Railroad

Gary Sharp, Federal Resources

Chris Weber, Smiths Detection

Scott Russell, Baltimore County Fire Department

About the Author

Chris Hawley is a founding partner in Blackrock 3 Partners, Inc., a firm that provides incident management training and consulting. Previously he was a deputy project manager for Computer Sciences Corporation (CSC) and was responsible for several WMD courses in the DOD/FBI/DHS International Counterproliferation Program. This program provides threat assessment, hazardous materials, and antiterrorism training with full-scale exercises worldwide. So far, Chris has provided training in more than 45 countries. Before his international work, Chris retired as a fire specialist with the Baltimore County Fire Department. Prior to this assignment, he was assigned as the special operations coordinator, reporting to the division chief of special operations and responsible for the coordination of the hazardous materials response team and the advanced technical rescue team, along with two team leaders. He was also assigned to the Fire Rescue Academy as one of four shift instructors. As a shift instructor, he was responsible for all countywide training on his shift. He has been a hazardous materials responder for over 19 years. Chris has 25 years' experience in the fire service, and prior to working in Baltimore County,

he was a hazardous response specialist with the City of Durham, North Carolina, Fire Department.

Chris has designed innovative programs in hazardous materials and antiterrorism and has assisted in the development of many other training programs. He has assisted in the development of programs provided by the National Fire Academy, Federal Bureau of Investigation, U.S. Secret Service, and many others. Chris has presented at numerous local, national, and international conferences, and he writes articles on a regular basis. He serves on the NFPA 472 committee and other pivotal committees and groups at the local, state, and federal levels. He also works with local, state, and federal committees and task forces related to hazardous materials, safety, and terrorism. Chris has published eight texts on hazardous materials and terrorism response. He has coauthored a text, *Special Operations: Response to Terrorism and HazMat Crimes*, along with Mike Hildebrand and Greg Noll, through Red Hat Publishing. He also assists a variety of publishers with the review and development of emergency services texts and publications.

Role of Monitoring in Hazardous Materials Response

OUTLINE

LEARNING OBJECTIVES

Upon completion of this chapter, you should be able to:

- Identify the use of monitors in emergency response
- Describe the standards and regulations involved with monitoring
- Describe applicable chemical and physical properties
- Identify the importance of vapor pressure in chemical emergency response
- Describe appropriate formulas and conversions

Hazardous Materials Alarm

Drums and Food on I-95—A truck carrying food rear-ended another tractor-trailer stopped in the middle lane of the busy interstate highway. It is estimated that the food truck was traveling at 65 mi/h (105 km/h) when it hit the stopped truck, which was carrying a mixed load consisting of mostly 55-gallon (208-liter) drums (Figure 1-2). The driver of the stopped truck was not hurt and was able to remove his shipping papers. The driver of the food truck was alive but significantly pinned in the wreckage. He was conscious and alert, and his initial vital signs were stable. The trucks were entangled, and the rear door to the truck carrying the drums was torn away. Drums had shifted onto the food truck and were lying over the front of that truck. The first responders saw the placards on the first truck and requested a hazardous materials assignment. When they approached the truck to evaluate the driver, they saw the drums in a precarious position in the back of the truck. They already had turnout gear on but also donned self-contained breathing apparatus (SCBA). They obtained the shipping papers from the driver and consulted with the hazardous materials company, which was about 20 minutes away. The drums contained mostly flammable solvents and some combustible liquids. The first responders were advised to use monitors, continue with full personal protective equipment (PPE), establish foam lines, and begin the rescue. Monitoring, a vital form of protection for the rescuers, was continuous. Upon arrival, the hazardous materials company met with the incident commander (IC) to evaluate the scene. They confirmed the monitoring done by the first responders and began to evaluate the other parts of the load. It was determined that whenever the rescue companies would move a part of the dash of the truck, one of the leaking drums would increase its flow. The hazardous materials company secured the leak, moved the drum, and secured the remainder of the drums. They examined the rest of the load to make sure there were no other problems, monitored the atmosphere, and observed that most of the load had been deformed by the impact but that there were no further leaks. Once the victim was removed and the rescue companies moved away, the hazardous materials team overpacked and pumped the contents of the other drums into other containers. The risk category was fire, and the PPE chosen was appropriate for the risk category. At no time were any flammable readings indicated during the rescue, although some were encountered during the transfer operation. This example illustrates how participants from various disciplines worked together to rescue a victim in a hazardous situation and how monitors provided a high level of protection for all.

1. What risks would the rescuers face if detection devices were not used?

2. What would you change about the response if flammable readings would have been found?

3. What level of flammable readings would present a risk?

Introduction

In the response to a variety of incidents, such as chemical incidents, confined space incidents, and other potential toxic environments, monitoring is one of the primary mechanisms for keeping emergency responders alive and healthy. With the threat of terrorist attacks or other criminal activity, monitoring and sampling play a major role. This task typically falls to hazardous materials teams, which are unique groups typically in the firefighting service that provide valuable assistance to fire fighters and the community (FIGURE 1-1). In some cases, hazardous materials teams take risks for the protection of life and property extending beyond a building, home, or vehicle to include water, air, land, trees, and the general environment. Hazardous materials

FIGURE 1-1 The Seattle, WA, hazardous materials team getting set up to mitigate an ammonia release.
Courtesy of Christopher Hawley.

teams may be requested because other fire fighters or law enforcement officers have encountered a situation that is outside their scope of expertise, that has endangered or injured them, or that presents a risk they are uncomfortable in dealing with. Because hazardous materials teams face unique challenges, monitoring plays an important role in keeping team members safe and should factor into every response.

In an emergency response, it is necessary to identify the risk the material present, make the situation safer, and then, under less stressful conditions, return it to normal. A **risk-based response (RBR)** philosophy focuses on the immediate fire, corrosive, toxic, and radioactive hazards requiring protective clothing, yet some forms of protection, such as a fully encapsulated gas tight ensemble, may be excessively protective and may create a more harmful situation than it is intended to guard against, creating risks resulting from heat stress or from slip, trip, or fall hazards.

Monitoring enables responders to assess and protect themselves against unidentified or unexpected situations and the related risks. For example, a common response-related question is what happens when a truckload of chemicals is damaged and the chemicals are mixed? Unless the materials react violently with each other, there is little concern if effective monitoring is performed. Usually, any violent reactions occur prior to arrival of the responders, and, if other reactions are

going to take place, there is usually sufficient visual or detected warning for the responders. Effective use of monitors identifies any potential risks for the responders, no matter the mixture.

According to risk-based response philosophy, the exact identity of the material does not matter. The RBR focuses on the immediate threat. The use of monitors allows you to minimize risk while saving lives and property, as you were sworn to do. This lifesaving ability is demonstrated in the Hazardous Materials Alarm box, "Drums and Food on I-95," and in **FIGURE 1-2**. A monitor takes all the gray area out of hazardous materials response and makes it black and white, if you can interpret what it is trying to tell you. When hazardous materials reference books tell responders to evacuate 7 miles (11 kilometers), and the monitors tell them to evacuate 7 feet (2.1 meters), which advisory is more accurate? What reflects real-time, real-life situations? Which distance would make the incident less complicated?

Unfortunately, monitoring is often not emphasized by hazardous materials responders as it should be, nor is it commonly used in the way it was intended. For many years, monitoring during an emergency response was an afterthought, but it is now considered essential for personnel protection. A hazardous materials response team that is not adept at monitoring is at an extreme disadvantage and can be placing its members and the public in harm's way. When searching for an unidentified chemical or dealing with multiple chemicals, responders may find monitoring and the new types of detectors overwhelming. Although the use of monitoring and sampling equipment requires practice, basic monitoring has become very simple (**FIGURE 1-3**).

FIGURE 1-2 Crews work to rescue the driver of the food truck on the right while hazardous materials crews remove drums of flammables from the other truck. Effective monitoring and risk-based response assisted in this rescue effort.
Courtesy of Baltimore County Fire Department.

SAFETY TIP

Monitoring protects responders from unidentified or unexpected situations. When you do not know the atmosphere, you must wear additional protection based on the perceived risk. This additional protective clothing adds risks resulting from heat stress or from slip, trip, and fall hazards, which may be greater dangers.

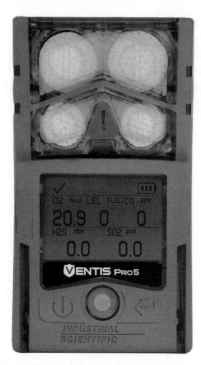

FIGURE 1-3 This device has the ability to detect 5 types of gases and will alert the user to dangerous situations.
Courtesy of Industrial Scientific.

FIGURE 1-4 It is essential for hazardous materials crews to check for fire, corrosive, toxic, and radiation hazards. This hazardous materials responder, in an encapsulating level A suit, is using a combination meter that has a photoionization detector, LEL, oxygen, CO, and H₂S sensors. He also has pH paper and a radiation pager as his initial entry instruments.
© Jones & Bartlett Learning. Photographed by Glenn E. Ellman.

SAFETY TIP

A response team that is not adept at air monitoring is at an extreme disadvantage and can be placing its members as well as the public in harm's way.

This book intends to provide the responder with detailed information related to the role of monitoring and monitoring devices in an easy-to-read format. Most fire service responders have a reasonable understanding of **flammable gas indicator**, but their knowledge about other detection technologies can be limited. Our society is becoming increasingly sophisticated, and our customers expect even better service. Technology is available to hazardous materials teams, and it can be readily purchased with technology that rivals a laboratory in capabilities. The downside of sophisticated devices is that some response teams rely too heavily on them and forget about the basics. There are a lot of basic detection devices that are low tech but can provide invaluable results. For example, Chapter 10 describes the use of iodine chips in the iodine chip test (HazCat), which is a simple test. Within a few seconds, we can identify the chemical family of an unknown liquid much faster than any expensive electronic devices, which is valuable information. Colorimetric tubes are another forgotten art, and they can provide valuable information about contaminants in the air, and more importantly, they tell the responders what is not present in the air.

The cost of a multiple gas detector and photoionization detector (PID) is now less than $3,000, well within the reach of most departments (**FIGURE 1-4**). The cost is quickly recouped by the amount of information and the level of safety that it furnishes. Because it is imperative that emergency responders understand what equipment is available and its limitations, this book presents information on monitoring, monitoring strategies, how monitors work, and their uses.

To truly understand monitoring, responders must be aware of the environment in which they work every day and should have basic knowledge in the many areas of hazardous materials response. With about 91 million chemicals listed by the **Chemical Abstracts Service (CAS)**, it is impossible to know everything about every chemical that exists. What are the most common responses for hazardous materials teams? What calls are run all the time? Which chemicals are being released every day? Once responders understand the fundamentals regarding monitoring, they can learn the chemical and physical properties, as well as the hazards, associated with the 20 most common chemicals involved in releases in the United States, which are listed in **TABLE 1-1** and **TABLE 1-2**. Once you learn the top 20 released chemicals, you have covered most of the chemical families that responders need to deal with. Effective use of monitoring protects responders from the very unusual events that occasionally occur.

TABLE 1-1 Top 20 Chemicals Released

Ammonia	Hydrochloric acid
Benzene	Hydrogen cyanide
Butadiene	Hydrogen sulfide
Chlorine	Methane
Combustible liquids (including various oils)	Methanol
Crude oil	Propane
Ethyl alcohol	Refrigerant gases
Ethylene glycol	Sodium hydroxide
Flammable liquids	Sodium hypochlorite
Gasoline	Sulfuric acid

Note: This list was taken from a consolidated list of many different top 10 lists provided by the Environmental Protection Agency (EPA), [Comprehensive Environmental Response Compensation and Liability Act (CERCLA), Emergency Reporting and Notification System (EIRNS), and the National Response Center (NRC) data] and the report to Congress under the Clean Air Act Amendments (112 r).
© Jones & Bartlett Learning

TABLE 1-2 Top 10 Bulk Chemicals Released During Transportation

Flammable liquids	Methanol
Corrosive liquids (acids)	Acetone
Sodium hydroxide	Caustic liquids (bases)
Alcohols	Potassium hydroxide
Hydrogen peroxide	Ethanol

From Department of Transportation (DOT) (reported spills, fatalities, injuries) 2017.

Regulations and Standards

Within the **Occupational Safety and Health Administration (OSHA) Hazardous Waste Operations and Emergency Response (HAZWOPER)** regulation (29 CFR 1910.120), there are not many specific requirements for monitoring. Similarly, the **National Fire Protection Association (NFPA)** Standard 472-2018 also has some requirements for monitoring but, like the OSHA regulation, is generic. Fundamentally, the responder must be able to use the detection equipment furnished by the authority having jurisdiction (AHJ) and use it according to the manufacturer's recommendations.

Both OSHA and the NFPA require a responder to be able to characterize an unidentified material. The HAZWOPER document includes monitoring information in its first part, which applies to hazardous waste sites. The emergency response section (paragraph q) barely mentions monitoring. The intention of 1910.120 is to protect workers, in this case the responders. OSHA requires the incident commander (IC) to identify and classify the hazards present at a site, and monitoring is the primary way to fulfill this obligation. Once an effective monitoring strategy is developed, the IC only has to wait for the results. If the monitoring shows little or no readings, the incident may be of small significance and limited risk, but if the monitoring results in high readings, the risk increases, and the incident shifts to a different level. High reading results can assist with PPE decisions and help the IC determine the scope of the incident and weigh public protection options.

LISTEN UP!

Both OSHA and the NFPA require you as a responder to be able to characterize an unidentified material.

LISTEN UP!

OSHA requires the IC to identify and classify the hazards present at a site.

The NFPA standard having the greatest impact on monitoring is the NFPA 472-2018 Standard for Professional Competence of Responders to Hazardous Materials/Weapons of Mass Destruction Incidents. NFPA 472 provides basic objectives related to monitoring, including knowing the appropriate equipment and the need for operational checks and calibration. It adds several monitoring objectives to Chapter 5 Hazardous Materials Operations section. In addition, the 2018 edition of NFPA 472 has mission-specific competencies for operations level responders in Chapter 6 that have monitoring and detection aspects. Section 6.7 is Mission-Specific Competencies: Detection, Monitoring, and Sampling. This section outlines the objectives for an operations level responder in the use of monitors.

In the objectives for the hazardous materials technician in Chapter 7, the NFPA requires a technician to be able to survey the hazardous materials incident to identify special containers, to classify unidentified materials, and to verify the presence and concentrations of hazardous materials using basic monitoring equipment. Several competencies at the technician level were moved, and Chapter 7 only requires the knowledge and skills of devices that can provide basic characterization of an atmosphere or material. The 2018 version of 472 added Chapter 19, which is Competency for Hazardous Materials Technicians with an Advanced Monitoring and Detection Specialty. Several objectives that used to be a technician skill are now an advanced skill. Another NFPA objective is to have the technician interpret the data collected from monitoring equipment and to estimate the extent of an endangered area. The technician must also know the steps for classifying unidentified materials. Being familiar with air monitoring is crucial to fulfilling these goals.

General Terminology

To be able to understand the remainder of this book, you must be familiar with general monitoring and chemistry terminology. These terms apply to all air monitors and detection devices unless specifically noted. Later chapters detail specific information about each of the instruments. The suggested readings section references texts that have further information on these and other terms.

State of Matter

The effective use of a detection device is determined by several factors, one of which is the state of matter the chemical is in. Most electronic detection devices are designed as vapor detection devices. The notable exceptions are those for biological, explosive, and radiological detection. Chemicals exist in three states of matter: Solids, liquids, and gases. Because science is not black and white, there are some variations of each of these. The severity of the incident can be determined by knowing if the material is a solid chunk of material, a pool of liquid, or an invisible gas. Of the 118 elements in the periodic table at room temperature, two are liquid (mercury and bromine), 11 are gases, 90 are solids, and 15 are unknown.

LISTEN UP!

The level of concern rises with each change of state; relatively speaking, a release of a solid material is much easier to handle than a liquid release, and it is nearly impossible to control a gas, but regarding air monitoring, it is much easier to detect gases as opposed to solids. See Table 1-2 for a summary of the interrelationships of chemical and physical properties.

When dealing with a release, the control methodology for each increases in difficulty from simple controls with a solid to difficult with a gas. Evacuation distances must be enlarged for releases involving gases, where a minimal evacuation distance would be required for a solid material (except for explosives). The route of entry into the body for chemicals that can harm humans also varies with the state of material. Solids usually can only enter the body through contact or ingestion, although inhalation of dusts is possible. Solids can take several forms, such as solid blocks, smaller chunks of material, powders, and even microscopic-sized solids that can float through the air. Alpha and beta emitters can be inhaled. For example, plutonium dust—an alpha emitter—can be inhaled and is extremely dangerous because it is now emitting alpha particles, which are 100 percent damaging to the lungs since none have enough energy to escape the body. Beta particles are extremely fast-moving electrons that react with the first

TABLE 1-3 Chemical Property Interrelationship				
If the chemical has:	**It then has a low:**	**It also has a high:**	**It has a wide:**	**It has a narrow:**
Low molecular weight	Boiling point Flash point Heat output	Vapor pressure Evaporation rate Ignition temperature	Flammable range	
High molecular weight	Vapor pressure Evaporation rate Ignition temperature	Boiling point Flash point Heat output		Flammable range

Chapter 1 Role of Monitoring in Hazardous Materials Response 7

matter they meet—this happens within feet for gases and much, much closer for liquids and solids. Although rare, some solids evaporate, changing their state of matter.

Liquids can be ingested, absorbed through the skin, and, if evaporating, inhaled. The rate of a liquid evaporating is based on the materials vapor pressure. A solid that has been mixed with a liquid is known as a slurry. The amount of movement that a liquid can undertake is described as viscosity. The more viscous a liquid is the slower it moves, and it could also be described as "thick." Compare water and ketchup, with the latter being more viscous. New motor oil is more viscous than used motor oil. Viscosity also applies to the other states of matter and is temperature dependent.

Gases and vapors can be inhaled, on occasion absorbed through the skin, and, to some extent, ingested. For a hazardous materials responder, the terms "gas" and "vapor" are synonymous, but to a chemist there are some differences. A gas is an element that is naturally occurring in that state. A vapor is a chemical that has changed its naturally occurring state of matter.

There are other terms used to describe gas-like substances, such as aerosol and fume. An aerosol is a solid or a liquid that is forced from a container by a gas. Spray paint is an example of an aerosol, which is a liquid material that is forced from a container by a gas, such as propane or nitrogen. Whipped cream uses nitrous oxide as a propellant, which is known as laughing gas and is a narcotic substance. A fume is a smoke-like substance that contains solid materials. Welding fumes can contain zinc, cadmium, beryllium, mercury, lead, and other compounds.

Melting and Freezing Point

The melting point (MP) is the temperature at which a solid must be heated to transform the solid to the liquid state. Ice has a melting point of 32°F (0°C). The freezing point is the temperature of a liquid when it is transformed into a solid. For water, the freezing point is 32°F (0°C), which may seem confusing as it is also the melting point. The actual temperatures vary by tenths of a degree but are very close.

Boiling Point

The **boiling point (BP)** is when the liquid is heated to a temperature where evaporation takes place and the liquid is changed into a gas (**FIGURE 1-5**). Another way to describe it is that evaporation is the process of a liquid changing to a gas and that happens below the boiling point. (Water evaporates at room temperature

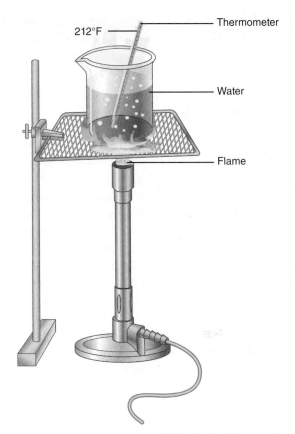

FIGURE 1-5 The point at which the liquid turns into a gas is called the boiling point.
© Jones & Bartlett Learning

and produces water vapor.) The boiling point is the temperature at which a liquid evaporates at its fastest rate. When a liquid boils, its vapor pressure equals the atmospheric pressure. It does not exceed the atmospheric pressure because evaporation takes energy, constantly reducing the temperature of the liquid to the boiling point, which can also be defined as the temperature of a liquid when its vapor pressure equals the atmospheric pressure. In reference materials, the listed boiling point is generally the temperature at which the vapor pressure equals 760 mmHg. Water boils at 212°F (100°C) and rapidly changes into a gaseous state. It is interesting to note that some boiling points are cold. Propane, for example, boils at −44°F (−42°C). The important thing to remember about boiling points is the fact that when the liquid approaches this temperature, vapors are being rapidly produced, which can cause serious problems.

LISTEN UP!

The NFPA requires a technician to be able to survey the hazardous materials incident to identify special containers, to identify or classify unidentified materials, and to verify the presence and concentrations of hazardous materials using monitoring equipment.

Vapor Density

Vapor density (VD) is the relative comparison of the weight of a material to air, with air having a value of 1. The vapor density determines whether the vapors rise or sink (**FIGURE 1-6**). The comparison comes from the **molecular weight** of the gaseous materials. (Molecular weight is the weight of the molecule based on the periodic table, or the weight of the compound when component atomic weights are combined.) Air has a molecular weight of 29, and materials with molecular weights greater than 29 are heavier than air and stay low to the ground. Those with molecular weights less than 29 rise in air. Vapor densities are calculated using molecular weight, dividing the molecular weight of a gas by the molecular weight of air. As an example, the molecular weight of methane is 16.04, which divided by 29 yields a vapor density of 0.55. Gases with a vapor density less than 1 rise in air; gases with a vapor density greater than 1 sink in comparison to air. This property is important because responders need to know where to sample and monitor to determine where the gases may be found. If the identity of the gas is unknown, first check low in the area being sampled, then midway, and then higher up. Even if you are sure what you are dealing with, it is a good practice to check, low, midway, and high.

Many reference texts provide vapor densities for gases and for some liquids that move to the gaseous state. The **National Institute of Occupational Safety and Health (NIOSH)** Pocket Guide lists the vapor density, as **relative gas density (RgasD)**, for all chemicals occurring as gases. For other materials and for chemicals for which the vapor density is not provided, compare the molecular weight of the gas with the molecular weight of air.

For practical purposes, most unidentified gases are heavier than air. Only 11 common gases rise in comparison to air. (See Table 1-4.) The mnemonic used to remember them is either DAMN2 3H CAE, or DAMN2 3H Cows Ate Everything. A former mnemonic for lighter-than-air gases, HAHAMICEN, left out hydrogen cyanide and diborane. Also, by molecular weight, hydrofluoric acid (HF) should be on the list, but with the addition of water vapor and the fact that HF binds with hydrogen, HF is almost two times heavier than air. HF does not act like its molecular weight unless it is very dry (low humidity) and it is very dilute. Chemicals with vapor densities close to the factor of 1 (including 1), such as 1.1, 1.2, 0.9, and 0.8, hang at midlevel and do not go anywhere unless there is wind to move them along; for example, carbon monoxide vapor density is = 0.97.

Weather affects vapor density and has a dramatic impact on how a gas behaves. The temperature of the liquid coming from a pressurized container also impacts where the vapors go. Liquefied ammonia is very cold when it is released from a pressurized cylinder and stays low until it heats up to 68°F (20°C), and then it will rise. The vapor densities provided in **TABLE 1-4** are at

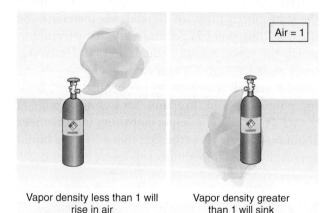

Vapor density less than 1 will rise in air

Vapor density greater than 1 will sink

FIGURE 1-6 In a normal, undisturbed atmosphere: **A.** Gases or vapors with a vapor density of less than 1 or with a molecular weight of less than 29 rise in comparison to air. **B.** Gases or vapors with a vapor density of greater than 1 or a molecular weight of more than 29 sink.

TABLE 1-4 Common Gases That Rise (Air = 1)			
Gas Name	**Vapor Density**	**Gas Name**	**Vapor Density**
Diborane (B_2H_6)	0.97	Ammonia (NH_3)	0.60
Methane (CH_4)	0.55	Neon (Ne)	0.7
Nitrogen	0.967	Hydrogen (H)	0.1
Helium (He)	0.138	Hydrogen cyanide (HCN)	0.9
Carbon monoxide (CO)	0.97	Acetylene (C_2H_2)	0.91
Ethylene (C_2H_4)	0.98		

standard temperature and pressure, as would be the case on an average day. Natural gas has a vapor density of 0.6, which means that it should go up. On days when the humidity is high or there is ground fog, natural gas tends to stay low. Propane, which has a vapor density of 1.6, is a real risk because it likes to stay low, but on hot dry days with lots of sun, it rises. These weather conditions are the extreme in most areas of the country, and when dealing outside the norm, do not expect chemicals to act in their usual fashion. This is another reminder that science is not black and white; there are considerable shades of gray (complexities) in the real world.

Vapor Pressure

By far, the most important physical property, not only to hazardous materials response but also to air monitoring, is **vapor pressure**. In Chapter 11 we discuss the importance of vapor pressure and risk-based response, and many of the issues discussed in this section are further explained in that chapter. We can define vapor pressure in several ways, but the best is related to the amount of vapor emitted from a material or how badly a solid or a liquid wants to become a vapor. Both solids and liquids have vapor pressure, although it most often is associated with liquids.

Another definition of vapor pressure is the amount of force applied on a specific container by the vapors coming from a material at a given temperature (**FIGURE 1-7**). Many responders measure vapor pressure by four common units of measurement: Millimeters of mercury (mmHg), pounds per square inch (psi), atmospheres (atm), and millibars (mb). Normal atmospheric pressure is equal to 760 mmHg, 14.7 psi, 1 atm, or 1,013 mb. As shown in Figure 1-7, to determine the vapor pressure of a material, it is placed in a sealed container that has a thermometer-like device in the cap. This device has a column of mercury in a cylinder that is graduated in millimeters. The force of the vapors pushes up on the column of mercury, and the highest level attained is the vapor pressure at the given temperature. Most temperatures are recorded at 68°F (20°C), which is considered

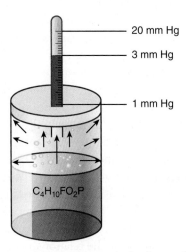

FIGURE 1-7 Vapor pressure is the force of the vapors pushing against the sides of the container. In this drawing the vapors push up the column of mercury, which corresponds to a reading of vapor pressure in millimeters of mercury (mmHg).
© Jones & Bartlett Learning

normal. We consider materials with a vapor pressure greater than 40 mmHg to be an inhalation hazard, or vapor hazard.

We can compare vapor pressures with water, which has a vapor pressure between 17 and 25 mmHg. Many responders use 25 mmHg because it is the highest recorded vapor pressure. When a spilled material has a vapor pressure less than 25 mmHg, it remains a liquid longer than the same amount of water. If a material has a higher vapor pressure than water, then it turns into a gas faster than the same amount of water. This is an important paragraph and one that should be applied in all hazardous materials responses. When materials change their state of matter and become airborne, their risk to humans increases. Vapor pressure is one aspect of this transformation. When materials have a vapor pressure of less than water, they do not readily get into the air. They may be toxic but are a contact hazard. When their vapor pressure is more than water, then the risk increases.

A third definition for vapor pressure is the most scientific: It is the force of the vapors emitted from the surface of the material at a given temperature, in terms of their ability to overcome the atmospheric force that keeps them in their current state of matter. Vapor pressure is the force of the material that comes off a substance and is related to another term, **volatility**. Often these terms are mistakenly used interchangeably. Volatility is the relative quantity of the vapors that come from the material, in terms of a specific standard (normally) at a specific temperature. Volatility is expressed in milligrams per meter cubed (mg/m^3), which can be converted to parts per million (ppm) using the formula found in Table 1-4. In most cases, the relationship is consistent because most high vapor pressure materials also have high volatility, but not all materials. For example, xylene has a low vapor pressure (9 mmHg) but is a flammable liquid that evaporates quickly because its volatility is somewhat high ($39,012 \ mg/m^3$), and that is due to its molecular structure. The chemical structure is a benzene ring and highly unsaturated, which becomes a major factor in its acting as a flammable liquid. Table 1-2 provides additional examples of chemical and physical properties.

Several other important factors related to vapor pressure and volatility are temperature, wind, and surface

type (**FIGURE 1-8**). The figures usually provided for vapor pressure or volatility are standard at 68°F (20°C), unless indicated otherwise. If local temperatures differ, the material reacts in relation to the temperature: The cooler the temperatures are, the less its vapors are an issue; the hotter the temperature is, the more vapors are produced. Air movement, or wind, also plays a role because increased air movement results in quicker evaporation transition to the gaseous state. In addition to air movement, the technician must also consider the vapor density of the product and its ability to be carried up and away. The last factor is the type of surface: Chemicals evaporate differently from different surfaces, such as concrete and sand. The surface area of the liquid does not have any impact on the vapor pressure, but it does impact the amount of vapors being released. If one container has a surface area of 3 square feet (0.28 m²) and another container has one of 9 square feet (0.84 m²), the larger one will produce a larger quantity of vapor.

TABLE 1-5 Common Conversions and Formulas	
Conversion or Formula Name	**Conversion or Formula**
Celsius to Fahrenheit	(°C $\times$ 1.8) + 32 = °F
Fahrenheit to Celsius	(°F − 32) ÷ 1.8 = °C
Percent volume to ppm	1% volume = 10,000 ppm
ppm to mg/m^3	$mg/m^3 = ppm \times \dfrac{\text{molecular weight}}{24.5}$
mg/m^3 to ppm	$ppm = \dfrac{(mg/m^3)\,(24.5)}{\text{molecular weight}}$
Vapor density	Molecular weight ÷ 29 = vapor density
Vapor density in high humidity	Molecular weight + 18 ÷ 29 = vapor density
Volatility	$V = \dfrac{16,020 \times \text{molecular weight} \times \text{vapor pressure}}{°\text{Kelvin}}$
°Kelvin	0°C = 273 Kelvin, so 68°F (20°C) is 293 Kelvin

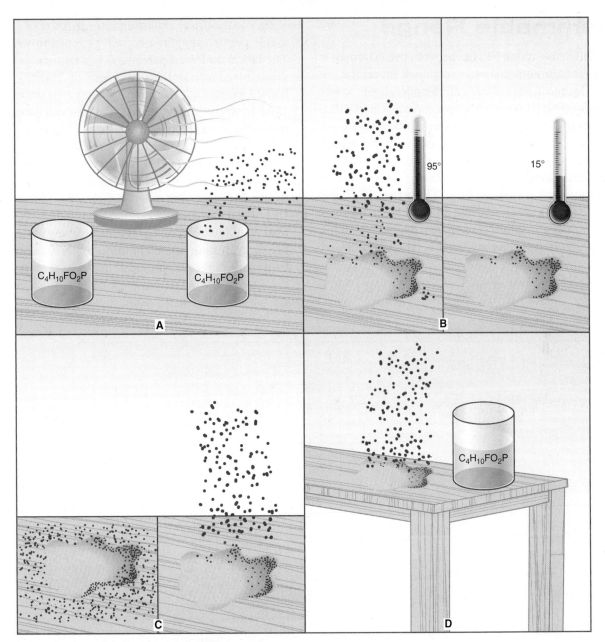

FIGURE 1-8 Vapor pressure and other factors play a role in how fast a chemical evaporates. **A.** Wind conditions. **B.** The temperature of the area around the spill and the temperature of the liquid. **C.** Type of surface. **D.** Surface area, size, and magnitude of spill.
© Jones & Bartlett Learning

LISTEN UP!

Volatility is the weight of the vapors that come from the material.

A material is considered an inhalation hazard if it has a vapor pressure greater than 40 mmHg, at which point the degree of risk presented by a chemical increases. Chemicals with a vapor pressure greater than 40 mmHg have greater potential to move away from the immediate site of the event, can be flammable, or require increased isolation and/or evacuation distances.

A chemical with a vapor pressure of 40 mmHg, whose vapors usually do not travel readily from the site of the release, usually create harm through touch or ingestion. They may be readily toxic or corrosive, but at the site of the release, they may also pose a potential inhalation hazard.

SAFETY TIP

A material is considered an inhalation hazard if it has a vapor pressure greater than 40 mmHg.

Flammable Range

The **flammable range (FR)** lies between two extremes: A minimum concentration and a maximum concentration. Typically, most responders call these levels the **lower explosive limit (LEL)** and the **upper explosive limit (UEL)**. Some texts refer to them as lower flammability limit (LFL) and upper flammability limit (UFL). No matter what the names are, the consequences are the same. The terms "LFL" and "UFL" refer to the air and fuel mixture for gases that can burn. The terms "LEL" and "UEL" describe the air and fuel mixture that ignites and then explodes. In either case, there is a rapid and quick fire, which in less than a second is typically followed by an explosion.

Key to the fire or explosion is the mixture of a flammable gas or vapor in a certain proportion with air. The LEL is the lowest percentage of a flammable gas or vapor mixed with air that can be ignited. The UEL is the highest amount of a flammable gas or vapor mixed with ambient air that can be ignited. Too little of a flammable makes the mixture too lean, and it cannot be ignited. Too much of a flammable makes for too rich a mixture, and a fire is not possible either. To have a fire or explosion, the concentration of the flammable material must be within the flammable range, between the LEL and the UEL (**FIGURE 1-9**). Flammable gas indicators (FGI) read up to the LEL point and indicate the possibility of fire or explosion. Some common materials and their flammable ranges are provided in **TABLE 1-6**.

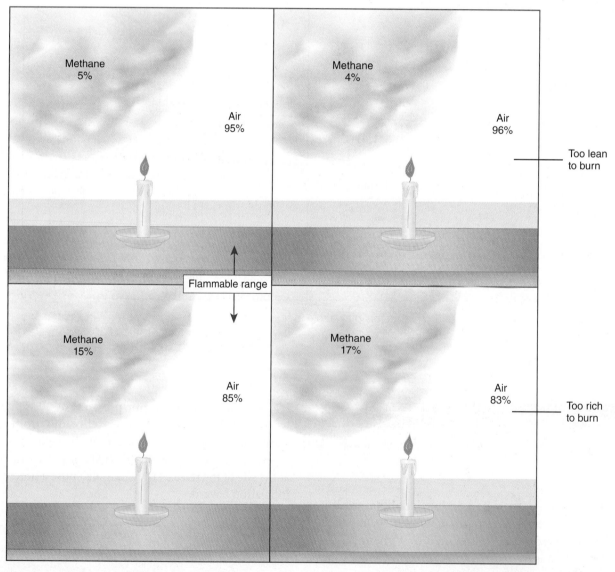

FIGURE 1-9 The flammable range is the range in which there can be a fire or explosion. Below the LEL or above the UEL there cannot be a fire.

© Jones & Bartlett Learning

The LEL is the lowest percentage of a flammable gas or vapor mixed with air that can be ignited. The UEL is the highest amount of a flammable gas or vapor mixed with air that can be ignited. Both require an ignition source.

TABLE 1-6 Flammable Ranges of Common Materials

Name	LEL (%)	UEL (%)
Acetylene	2.5	100
Acetone	2.5	12.8
Ammonia	15	28
Arsine	5.1	78
MEK (methyl ethyl ketone)	1.4 @ 200°F (93°C)	11.4 @ 200°F (93°C)
Carbon monoxide	12.5	74
Ethylene oxide	3	100
Gasoline	1.4	7.6
Hydrazine and UDMH (unsymmetrical dimethyl hydrazine)	2.9 (2%)	98 (95%)
Kerosene	0.7	5
Toluene	1.1	7.1
Hexane	1.1	7.5
Propane	2.1	9.5

© Jones & Bartlett Learning

Flash Point

Associated with the flammable range is the **flash point (FP)** of a material. The flash point is the temperature of a liquid at which the liquid gives off sufficient vapors that, when mixed with air, puts it in the flammable range and makes it capable of being "flashed" (momentarily ignited) by an ignition source. There are three keys to this definition:

1. The temperature of the liquid at which vapors can be given off,
2. The vapors mixing with air to reach the flammable range,
3. An ignition source.

The flash fire can be duplicated only in the lab; in the real world, when something flashes, it generally continues to burn, thereby reaching its **fire point**. In the street, the two temperatures are so close that a quick flash fire is not likely.

Accuracy and Precision

Two terms with different characteristics are commonly used to describe features of monitors: **Accuracy** and **precision (TABLE 1-7)**. Accuracy is the ability of the meter to produce findings as close as possible to the actual quantity of gas. An accurate meter, exposed to 80 ppm of a known gas, should read 80 ppm. A meter that is not accurate, exposed to 80 ppm, may read only 30 ppm. Precision describes the ability of the monitor to reproduce the same results each time it samples the same atmosphere. In other words, it duplicates the readings every time the same concentration of gas is sampled. If a monitor is exposed to 70 ppm of carbon monoxide, the monitor should display 70 ppm, or at least a close reading, for every sample. If the monitor displays 40 ppm with every sample, it is precise but not accurate. A meter

TABLE 1-7 Accuracy and Precision

Meter	Known Quantity of Gas	1st Reading	2nd Reading	3rd Reading	4th Reading	Meter Description
Meter A	100 ppm	50 ppm	50 ppm	50 ppm	50 ppm	Precise
Meter B	100 ppm	100 ppm	100 ppm	100 ppm	100 ppm	Precise and accurate
Meter C	100 ppm	70 ppm	90 ppm	85 ppm	95 ppm	More accurate than Meter A

© Jones & Bartlett Learning

that reads 64, 81, and 60 for the same 70-ppm sample is more accurate because it is providing values close to the actual sample concentration, but it is not precise due to the variation of the readings. "Precision" refers to the repeatability of the readings. An air monitor can be precise but inaccurate and vice versa. See Table 1-6 for more examples.

Many factors affect precision and accuracy, including, but not limited to, the chemical and physical properties of the sample, weather (humidity and temperature are the two biggest factors) and the sensor technology, the condition of the meter (age, maintenance, degree of use, etc.), and how the monitor is used (user error).

Bump Tests and Calibration

Four other terms that need explanation are **bump test**, calibration check, **calibration**, and fresh air calibration. When some sensors fail, they fail to the zero point, and in some cases, you may not know they failed. A calibration check is the optimal determination that the sensors are indeed functioning as intended, a quantitative test of the sensors using calibration gas (a known concentration). The bump test is also known as a field test, more of a qualitative test exposing a monitor to known gases and allowing the monitor to go into alarm, thus verifying the monitor's response to the gas. It is not expected that the readings exactly match those of the bump gas, but they should be close. Bump gas comes in small disposable cylinders, much like a spray can. For a four-gas mixture (pentane [LEL], O_2, CO, and H_2S), the cost is about $75 and has a shelf life of six months. An inexpensive method to bump test a monitor is to use a felt tip marker cap, but this method is far from scientific. Fresh air calibration is not calibration; it is resetting the sensors to a known zero point or baseline. The oxygen sensor will be set at 20.9 percent.

Calibration

Properly performed calibration confirms that your monitor responds accurately to a known quantity of gas. As an example, to calibrate a carbon monoxide (CO) sensor, you connect the CO sensor to a source of CO. The source of CO is a cylinder that contains 50 ppm of CO (**FIGURE 1-10**). When you calibrate a sensor, you input into the device that you will be exposing the sensor to 50 ppm CO, then when the sensor is exposed to the calibration gas it sets itself to read 50 ppm. The sensor is not actually interpreting what it sees as 50 ppm

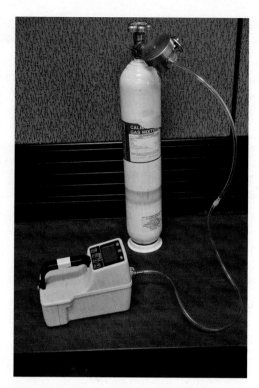

FIGURE 1-10 An example calibration setup in which a known quantity of gas is drawn across the sensors, and they are set to read the appropriate level of gas.
© Jones & Bartlett Learning. Photographed by Glenn E. Ellman.

of CO. It is simply outputting "raw" data caused by its exposure to the known 50 ppm concentration of CO, and you have told the meter that amount of "raw" data happens to be equal to 50 ppm of CO. New sensors usually read higher than they are intended; calibration (manual or automatic) electronically changes the sensor to read the intended value. As the sensor gets older, it becomes less sensitive, and calibration electronically raises the value that the sensor displays. A sensor that cannot be electronically brought up to the correct value is dead and needs to be replaced. Calibrations used to be very difficult, but with the new monitors and multigas cylinders, it is relatively easy, depending on the manufacturer. The regularity of calibration is subject to debate; some response teams calibrate daily, some monthly, and others every six months. Most of the written instruction guides from the manufacturers require calibration before each use. The definition of a calibration at this point is also subject to debate because we can verify the monitor's accuracy by exposing it to a known quantity of gas but not perform a "full" calibration. Most response teams establish a regular schedule of calibration (weekly or monthly) and then perform bump or field checks during an emergency response. Check with the manufacturer as to what calibration or bump test policy they recommend; they are all different. Calibration should be done at the temperature and atmospheric pressure that the device will be used. The frequency of calibration depends on several factors, but the principal factor is the frequency of use for the instrument. If the instruments are being exposed to a variety of gases on a regular basis, then the instrument should be calibrated more frequently. If the sensors are going to be exposed to materials that can damage or poison the sensors, calibration should be more frequent. If the sensors are overranged or exposed to a large amount gas or a corrosive gas, no matter the quantity, they should be calibrated. Any new sensor that is placed in a device should be calibrated. Any time there is a major change to the instrument, either intentional or unintentional, it should be calibrated. Always bump test the instrument to make sure it responds appropriately and will alarm.

> ### LISTEN UP!
>
> No matter the calibration frequency, all instruments should be bump tested prior to use to ensure the sensors react and alarm.

There are a variety of calibration gas combinations; some are single gas cylinders, and others are multigas cylinders. Calibration gas consisting of a four-gas mixture (LEL, O_2, CO, H_2S) costs about $150 and has a shelf life of about a year. Some gases such as chlorine and ammonia have shorter from shelf lives and are substantially more expensive. The tubing that you use to carry the gas also can impact the accuracy of the calibration. Always use the tubing and equipment provided by the manufacturer. Some tubing absorbs the calibration gas, which means the sensor may not be reading the exact quantity of gas that is being pushed across the sensor. Remember that you input into the device the amount of gas that the sensor will see. If you tell the sensor it should see 50 ppm CO, and less CO is received by the sensor, the sensor will be inaccurate since it will indicate 50 ppm when exposed to a lower concentration of CO. The concentration of the gas varies from manufacturer to manufacturer, but some manufacturers provide a higher concentration. Sensors are like batteries, and the more they work and are exposed to gases, the sooner they will die. Many of the manufacturers use 50 ppm CO to calibrate the CO sensor. If a manufacturer supplies 100 ppm CO for calibration, the sensor may have a shorter life. One would think that using the smallest concentration of a calibration gas would be advantageous, but unfortunately, this isn't true for all gases and sensors. There

is a margin of error when completing a calibration, and when you use a smaller concentration of gas, the margin of error can have more of an impact. When calibrating gases other than for LEL, CO, H_2S, and O_2, there may be some rest time required when calibrating other toxic sensors. Exposing the sensors in a multigas instrument to ammonia, as, an example, can have an impact on the other sensors. Calibrate the typical sensors, let the instrument run for a while, and let the calibration gases clear. Then calibrate the ammonia sensor, and if there are other sensors that need calibration, allow another rest period before starting the next calibration. One other test to be conducted during the bump test is a pump check. The user should follow the manufacturer's recommendation on checking the validity of the pump. If the pump fails, the device is out of service.

Bump Tests

The best recommendation is to always bump test your monitors, use one instrument while performing a full calibration on the second unit, and then switch to verify the results of the first monitor (**FIGURE 1-11**). If you do not calibrate your instruments prior to an event, then immediately after the event calibrate them to ensure their accuracy and precision.

> ### LISTEN UP!
>
> One of the reasons regarding the confusion with calibration is the lack of regulatory guidance on the issue. OSHA does not have a clear requirement to calibrate instruments except to follow the manufacturer's recommendations. The International Safety Equipment Association (ISEA) is trying to have OSHA adopt its recommendations for calibration. Many air monitoring manufacturers have adopted this philosophy as well. ISEA recommends that the instruments receive a functional (bump) test prior to each day's use, using a known concentration of gas, which can demonstrate an appropriate response. If the instrument fails the functional test, it must receive a full calibration. If the instrument is exposed to damaging environmental conditions that may damage the sensors, tests should be more frequent. ISEA allows for less frequent checks but primarily with regard to a fixed facility, where there are known hazards, and the instrument is assigned to one worker who tracks the use of the instrument with an equipment use log. In this case, the instrument must be checked for the initial 10 days in the intended atmosphere. The sensors should be exposed to all potential hazards for a sufficient time. If the daily checks confirm that it is not necessary to make adjustments, then the time between checks can be up to 30 days.

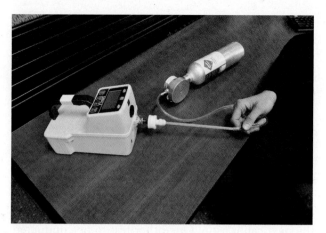

FIGURE 1-11 This instrument is being bump tested. A known quantity of gas in being drawn into the instrument. The instrument should respond to the gas and alarm, which accomplishes the bump test.
© Jones & Bartlett Learning. Photographed by Glenn E. Ellman.

Fresh Air Calibration

Many instruments when they are switched on will ask the question—"Fresh air calibration?" and if you know you are in clean air free of contaminants, you can answer yes. If you are not sure about the air quality, then press no. Fresh air calibration is not calibration; it is the resetting of the sensors to read zero and the oxygen sensor to read 20.9 percent. You may not reset the sensors to read zero. The only way to truly zero the sensors is to use a zero-gas cylinder and calibrate using zero gas, which is free of contaminants. Some manufacturers also offer charcoal traps to assist in a more accurate fresh air calibration. When using any gas detection device, the monitor should remain on in clean air until it is free of any contaminants. If a device is used to detect CO and the device is reading 100 ppm, and the device is shut off while the detection device is reading 100 ppm, there is some amount of CO that remains in the sensor. When the device is turned back on, the CO is still in the sensor, and if the startup series is allowed, the sensor will read the amount remaining in the sensor. If the user conducts a fresh air calibration with CO in the sensor, the sensor will zero out electronically. The CO remains in the sensor, and as it clears out, the reading will become negative. Detection devices cannot read

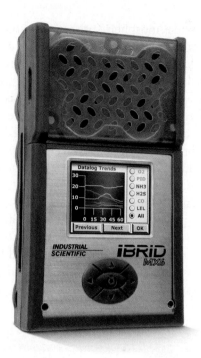

FIGURE 1-12 This device has 6 sensors and has an advanced side and can datalog the exposure the crews are receiving.
Courtesy of Industrial Scientific.

in a negative mode and so will generally alarm or go in fault mode, and in most cases, the screen will indicate "NEG" to show the sensor is reading in negative numbers. To clear this error, the device must run until the CO clears the sensor, and then the fresh air calibration is run again zeroing the sensor.

Using a monitor that logs data internally assists in the record keeping necessary to protect yourself in court (**FIGURE 1-12**). Another method of providing protection is to keep good calibration records. If you compile regular calibration records, you can determine how much your monitor drifts from calibration to calibration. Once you have several months' worth of data, you can be reasonably certain as to how accurate and precise your monitor is. With these records, you can also determine the useful life of your sensors and can anticipate when they will need to be replaced.

Response and Recovery Times

All monitors have a certain reaction time, or response time, which many manufacturers specify. The response time is the amount of time that the meter takes to report a reading that is 80–90 percent accurate. You will see these represented by "T" and an associated number. T^{50} 45s means the reading will be 50 percent accurate in 45 seconds. T^{90} means 90 percent accurate. These ratings are generally associated with radiation detectors, but on

---

LISTEN UP!

Never turn a detection device off with contaminating gas in the sensor; make sure the device is clear and reading zero. Any gas remaining can damage the sensor, tubing, and electronics and can impact the usability of the device when it is turned on the next time.

occasion, they are shown for other sensors. The reaction time varies with the type of sensor and whether you are sampling with a pump.

- Monitors without a pump operating in a diffusion mode generally have a 30- to 60-second lag time, not counting the sensor reaction time.
- Monitors with a pump have a typical reaction time of less than 20–30 seconds, but even this can vary among manufacturers.
- A photoionization sensor reacts in a few seconds, and an LEL sensor takes about 10 seconds to react.
- An oxygen sensor can take 20–30 seconds to react with a pump.
- Some toxic sensors have a reaction time of 20–60 seconds.
- Some detection devices for weapons of mass destruction (WMD) agents have reaction times of 90 seconds or more for low concentrations.

The goal is to provide a given amount of volume of air across the sensors. When using sampling tubing less than 1 foot (30 cm), add 1–2 seconds of reaction time for the extension. For longer extensions, less than 10 feet (3 m) of hose, you may need to add 4 seconds delay in addition to the sensor reaction time. Be sure to follow the manufacturer's recommended hose type, lengths, and diameter of hose to ensure that the pump operates correctly. Only use hose provided by the manufacturer because some hoses can absorb the chemicals you are sampling and cause problems in future sampling efforts. Teflon hose is the preferred hose, and

Extension Delay Times

Type/Size	Delay Time Per			
	1' (30 cm)	10' (3 m)	30' (9 m)	100' (30 m)
Teflon probe 1/8"	0.1	1	5	29
Teflon 0.15" (4 mm)	0.4	4	11	38
Teflon 1/4" (6 mm)	1.1	11	33	112

Excerpted from RAE by Honeywell Technical Note - 140.

Sensor Response Times

Sensor	Range	T50 Response (Seconds)	T90 Response (Seconds)	Notes
Oxygen electrochemical	0–30%	6	15	Accuracy varies 0.5–0.8% with differing temperatures and humidity
LEL – Catalytic bead LEL	0–100% LEL	15	35	Accuracy varies 5–15% with differing temperature and humidity
LEL – Infrared	0–100%	15	35	Accuracy varies 5–15% with differing temperature and humidity
Ammonia electrochemical	0–500 ppm	21	78	Accuracy varies up to 15% with differing temperature and humidity
CO	0–1500 ppm	8	18	Accuracy varies up to 15% with differing temperature and humidity
CO/H_2S low concentration	0–1000 ppm	9	20	Accuracy varies 5–15% with differing temperature and humidity
Hydrogen cyanide	0–30 ppm	25	80	Accuracy varies 10–15% with differing temperature and humidity
PID 10.6 eV	0–2000 ppm	15	20	Accuracy varies 10–15% with differing temperature and humidity

For ISC MX6 iBrid multigas detection device.
© Jones & Bartlett Learning

Tygon tubing should not be used. Tygon absorbs some flammable materials, reducing the amount of the gas reaching the sensor.

Monitors also have a **recovery time**. Recovery time is the amount of time that it takes the monitor to clear itself of the air sample. This time is affected by the chemical and physical properties of the sample, the length of sampling hose, and the amount absorbed by the monitor. If a CO sensor takes 30 seconds to read 100 ppm, once you remove the monitor from the contaminated environment, it will take at least 30 seconds to clear the sensor, but if the device is in a contaminated environment that consists of CO for 5 minutes, it will take several minutes for the sensor to clear.

Correction Factor

Monitors are calibrated to a specific gas. Some LEL monitors are calibrated for methane (using pentane calibration gas) and are accurate and precise for only methane. Other monitors are calibrated using a pentane scale, using methane as the calibration gas. The **correction factor**, also known as a **relative response factor**, is the way the monitor reacts to a gas other than the one it was calibrated for. The monitor's manufacturer tests the monitor against other gases and provides this factor, which can be used to determine the amount of gas present when sampling. Some of these factors are provided in **TABLE 1-8**. For example, suppose at a hexane spill you are using an Industrial Scientific Corporation (ISC) iBrid MX6, calibrated for pentane, and the detector is reading 68 percent of the LEL. The correction factor for hexane is 1.42, so you multiply this factor by the LEL reading.

$$\text{Detector reading} \times \text{correction factor} = \text{actual LEL reading}$$
$$68 \times 1.42 = 97 \text{ (rounded up)}$$

The actual LEL is 97 percent, a situation that is worse than what the instrument indicates. If you used the same instrument for a release of acetylene, the correction factor is 0.7. Assuming again that the instrument is reading 68 percent of the LEL, the actual reading is 48 percent ($68 \times 0.7 = 48$), a safer situation than reported by the detector. It is not important to memorize response factors, but it is a good idea to have the correction factors for the chemicals you commonly respond to listed on a laminated card attached to the monitor or in the storage case.

The important correction factors are those greater than 1, which means that the atmosphere is worse than what your monitor is reading. If dealing with legal issues is a possibility, it is important to do your math and calculate the correction factors for the report. If the manufacturer does not provide a correction factor for the chemical you are dealing with, you can either contact the manufacturer or estimate the factor using molecular weights. For the most part, manufacturers' correction factors follow the molecular weight of the chemicals and are the most accurate. Your use of molecular weight may not be not exact, but it can get you into the range of the factor. Table 1-8 shows that the response curve factors of the Industrial Scientific iBrid MX6 track closely to the molecular weights, as do the MSA (mine safety appliances) factors.

SAFETY TIP

The important correction factors are those greater than 1, which means that the atmosphere is worse than what your monitor is reading.

LISTEN UP!

Correction factors are known by several names, which can be confusing. Some of the other names are relative response and response curve. The term "correction factor" is the best one because it correctly describes the action taken; you are correcting the meter reading using a factor supplied by the manufacturer.

Formulas and Conversions

Several formulas and conversions apply in air monitoring. **TABLE 1-9** provides some of the chemical and physical properties of a handful of chemicals. If you can identify at least one chemical and physical property, you can use this chart to compare it to other known chemicals. Gasoline and fuel oil 2 (diesel fuel) provide a good comparison for this chart. Table 1-2 provides the interrelationships of some of the chemical properties. Table 1-4 provides some commonly used conversions and formulas.

TABLE 1-8 Correction Factors (Calibrated with Methane)[a]

Gas Being Sampled	ISC iBrid MX6 Factors[b]	MSA ALTAIR 5X Factors[c] (Pentane Gas)	QRAE 3 (RAE System Factors[d])	Molecular Weight
Hydrogen	0.94	0.5	1	1.0
Methane	1	0.5	1	16.04
Acetylene	1.3	0.6	2.9	26
Ethylene	1.3	0.5		28.06
Ethane	1.3	0.6	1.4	30.08
Methanol	1.1	0.5	1.6	32.1
Propane	1.62	0.6	1.4	44.1
Ethanol	1.49	0.6	1.8	46.1
Acetone	1.7	N/A	1.9	58.1
Butane	1.67	0.7	~1.9	58.1
Isopropanol	1.9	0.9	2.2	60.1
Pentane	2.02	0.7	~2	72.2
Benzene	1.9	1.1	2.1	78.1
Hexane	2.86	0.9	2.1	86.2
Toluene	2.55	1.0	2.47	92.1
Styrene	2.2	N/A	2.4	104.2
Xylene	2.5	2.5	~2.8	106

a – Always use the correction factors supplied by the manufacturer and for the specific sensor, and keep in mind these are laboratory estimations.
b – Factors are for an Industrial Scientific MX6 iBrid calibrated with methane.
c – Factors are for a MSA ALTAIR 5X LEL Sensor calibrated with pentane simulant.
d – Factors are for QRAE 3 – RAE Systems by Honeywell calibrated with pentane.
© Jones & Bartlett Learning

TABLE 1-9 Chemical and Physical Properties of Selected Chemicals

Chemical	Molecular Weight	Freezing Point (°F/°C)	Boiling Point (°F/°C)	Flash Point (°F/°C)	Vapor Pressure (mmHg)	Autoignition Temperature (in Air) (°F/°C)	LEL (%)	UEL (%)	Ionization Potential (eV)
Acetone	58.1	−140/−96	133/56	0/−17	180	869/465	2.5	12.8	
Acetylene	26	−119/−84	Sub.	Gas	33,440	581/305	2.5	100	11.4
Ammonia	17	−108/−78	−28/−33	Gas	6460	1204/651	15	28	10.18
Arsine	78	−179/−117	−81/−63	Gas	11,400		5.1	78	9.89
1,1dimethylhydrazine	60.1	−72/−58	147/64	5/−15	103	480/249	2	95	8.05
Dimethyl sulfate	126.1	−25/−32	370/188	182/83	0.1	370/188			<11.7
Ether	74.1	−177/−116	94/34	−49/−45	440	356/180	1.9	36	9.53
Ethion	384.5	10/−12	302/150	349/176	0.0000015				
Fuel oil #2	~170	−238/−150	347/175	126/52	5 (100°F/38° C)	494/257	0.7	5	<10.6
Gasoline	~72	−267/−166	102/39	−50/−46	300	880/471	1.4	7.6	<9.8
Hydrazine	32.1	36/2	236/113	99/37	10	74–518/ 23–270	2.9	98	8.93
MEK	72.1	−123/−86	175/79	16/−9	78	759/404	1.4	11.4	9.54
Toluene	92.1	−139/−95	232/111	40/4	21	896/480	1.1	7.1	8.82
Triethylamine	101.2	−175/−115	193/89	20/−7	54/12	480/249	1.2	8	7.5
Xylene	106.2	−13/−25	292/144	90/32	7/−14	867/464	0.9	6.7	8.56

After-Action REVIEW

IN SUMMARY

The basic terminology and chemical properties are key to understanding monitoring. The terms defined in this chapter are used in later chapters, so it is essential that you understand them. With the assistance of monitoring and with background knowledge of chemical properties, assessments can be made as to the type and severity of the event. All monitors are essentially dumb devices that take readings and output a number. It is up to the human using the device to interpret that number and make an educated assumption as to what it means. Chemical and physical properties are important in making that assumption and in the successful handling of the event.

KEY TERMS

Accuracy Term to describe how closely a monitor's readings are to the actual amount of gas present.

Boiling point Temperature at which liquid changes to gas. The closer the liquid is to the boiling point, the more vapors it produces.

Bump test Using a known quantity and type of gas to ensure that a monitor responds and alarms to the gases that are being tested.

Calibration Electronically adjusting the monitor to read the same as a calibration gas, which is an accurately known gas concentration.

Calibration check Checking the response of a monitor against known quantities of a sample gas (most commonly against calibration gas).

Chemical Abstracts Service (CAS) A service that registers chemical substances and issues them unique registration numbers, much like a Social Security number.

Correction factor Factor that applies to how a given gas reacts with a monitor calibrated to a different gas.

Fire point The lowest temperature at which a liquid ignites and, with an outside ignition source, continues burning.

Flammable gas indicator A gas monitor that is designed to measure the relative flammability of gases and to determine the percent of the lower explosive limit. Also known as LEL monitor, or combustible gas detector.

Flammable range The numeric range, between the lower explosive limit and the upper explosive limit, in which a vapor burns.

Flash point The minimum temperature of a liquid that produces sufficient vapors to form an ignitable mixture with air when an ignition source is present above the liquid.

Hazardous Waste Operations and Emergency Response (HAZWOPER) An OSHA regulation that covers waste site operations and response to chemical emergencies. Found in 29 CFR 1910.120.

LEL See Lower explosive limit.

Lower explosive limit (LEL) The lower limit of the flammable range; the lowest percentage of flammable gas and air mixture in which there can be a fire or explosion.

Molecular weight Weight of a molecule based on the periodic table or the weight of a compound when the atomic weights of the various components are combined.

National Fire Protection Association (NFPA) A consensus group that issues standards related to fire, hazardous materials, and other life safety concerns.

National Institute of Occupational Safety and Health (NIOSH) The research agency of OSHA that studies worker safety and health issues.

NFPA See National Fire Protection Association.

NIOSH See National Institute of Occupational Safety and Health.

Occupational Safety and Health Administration (OSHA) Government agency tasked with providing safety regulations for workers.

OSHA See Occupational Safety and Health Administration.

Precision The ability of a detector to repeat the results for a known atmosphere.

RBR See Risk-based response.

Recovery time The amount of time it takes for a detector to return to 0 after exposure to a gas.

Relative gas density (RgasD) Term used by the NIOSH Pocket Guide that means vapor density. It is a comparison of the weight of a gas to the weight of air.

Relative response factor See Correction factor.

Risk-based response A system for identifying the risk chemicals present even though their specific identity is unknown. This system categorizes all chemicals into fire, corrosive, or toxic risks.

UEL See Upper explosive limit.

Upper explosive limit (UEL) The upper limit of the flammable range; the maximum concentration of flammable gas or vapor that can be mixed with air to have a fire or explosion.

Vapor density (VD) The weight of a gas compared with an equal amount of air. Air is given a value of 1. Gases with a vapor density less than 1 rise, while those with a VD greater than 1 sink.

Vapor pressure The force of vapors coming from a liquid at a given temperature.

Volatility The amount of vapors coming from a liquid.

Review Questions

1. What are the four risk categories used in a risk-based response?

2. What is vapor pressure?

3. What is the difference between vapor pressure and volatility?

4. What is considered vapor pressure that may create inhalation hazards?

5. Most gases and vapors travel in which direction when released?

6. Define flash point.

7. What is the flammable range?

8. How often should detection devices be calibrated?

CHAPTER 2

Toxicology

LEARNING OBJECTIVES

Upon completion of this chapter, you should be able to:

- Describe the various types of exposures
- Describe the difference between acute and chronic exposures
- Identify the categories of hazards using TRACEM
- Describe the routes of exposure
- Describe the exposure levels and their use in response
- Describe the use of Acute Exposure Guidelines

Hazardous Materials Alarm

Garbage Truck Versus Van—A hazardous materials assignment was dispatched to an auto accident with entrapment, in which a dumpster-style garbage truck had been involved in a head-on collision with a van. The van was marked as a carpet-cleaning van and had two occupants who were well entangled in the wreckage. The forks had been down on the garbage truck, complicating the rescue. The rescue crews who were working to free the victims said the passenger was complaining of his face burning. The rescue crews, all in full personal protective equipment (PPE), looked in the back of the van where all the cleaning chemicals had been thrown around, many of them leaking. We figured that some of the cleaning chemicals had been splashed on the passenger. We checked the pH of the victim, which was severely acidic, while the pH of the mixture in the back of the van was on the opposite end of the scale, to the caustic side. We quickly removed the victim and began the decontamination process, taking care of his head and face. We continued to check the products in the back of the van, looking for the acid so we could let the hospital know what material had been spilled on the patient. Unable to locate any acids, we started looking around the front of the van and in the garbage truck. We determined that one of the forks from the garbage truck had gone through the battery on the passenger side of the van, splashing battery acid on that side of the van.

1. What type of PPE would be recommended?

2. What would be a high priority for patient treatment?

3. What would be a high priority for the rescuers?

Introduction

Like the chemistry information in Chapter 1, a hazardous materials responder proficient in monitoring needs to have a more than basic understanding of **toxicology** and should understand the science. One caution is that although toxicology is discussed as an absolute science, it could be considered a gray science. There are many factors that can impact a chemical's impact on human health. The largest variable is the human, as each one is different from the other. There is an understandable fear of being exposed to chemicals, and if uninformed, that fear may be unfounded or cause unnecessary worry. In the workplace, every chemical has a corresponding **safety data sheet (SDS)**, which is an OSHA requirement. These SDSs provide information on the chemical and physical properties, but most importantly, they provide information on the health hazards and toxicology-related information. For responders, there are a variety of texts and online references available. See the textbox on references for additional information. Toxicology is the study of poisons and their effect on the body, and people who study the effect of poisons on the body are known as toxicologists. Although toxicologists are typically found in the medical community, most industrial facilities have **industrial hygienists** on staff whose responsibility is to protect the workers' health and safety. Some fire departments have employed or contracted with industrial hygienists to help protect responders. These people have extensive training in toxicology and chemical exposures and are great resources to the emergency services. Because the world of toxic exposures can be complicated, and in emergency situations information is needed quickly, a quick consultation with an industrial hygienist may make the incident easier to resolve. When a toxicologist is not available, the SDS should provide information on the potential health hazards, signs and symptoms of exposure, and typically some response information. One prime piece of information is a phone number in which to call for additional information.

Types of Exposures

Being exposed to a chemical that is considered toxic may present a risk, the level of which is typically spoken of in terms of the chemical's potential *hazard*. Having an exposure to a chemical does not mean there will be an adverse impact. A toxic material is one that has potential to cause harm, but the potential harm is dependent on several possibilities. In common use, the phrase "the dose makes the poison" means that a certain dose can be hazardous. A material that is **highly toxic** requires a smaller dose to cause a reaction, as opposed to something that is mildly toxic. Aspirin (acetylsalicylic acid) is a common material that most people take on a regular basis without any issue (**FIGURE 2-1**). The usual dose for aspirin is 325 mg that is taken four times a day. The references vary but list a potentially lethal dose for a 200-pound (91 kg) person as being between 40 and 100 pills. Vitamin D (cholecalciferol) is a common vitamin supplement that is found in milk, shown in Figure 2-1, and is available in a variety of other forms. The level in milk averages 115–124 **international units** (IU) per 8 ounces (2.9–3.1 µg in 237 mL). The recommended dietary allowance (RDA) dose for 19- to 50-year-old persons is 15 µg (600 IU) per day. When levels exceeding 1250 IU (30 µg) per day for a period of weeks are ingested, health issues can arise. The published upper limit dose is 50 µg (2000 IU) per day. The one interesting fact about vitamin D is that it is also used as rat poison. The thought process is that rats eat the poison and within a few days or a week die from the toxic vitamin D effects. Even water can have toxic effects, as ingesting too much water alters the body's chemistry. In 2007, a radio station in Sacramento, CA, conducted a contest, which forced contestants to drink large amounts of water. One woman drank less than

FIGURE 2-1 Aspirin is a common material that most people take on a regular basis without any issue, but a relatively small number of pills can have a fatal effect. Vitamin D is a common vitamin supplement, which is found in milk and is available in pill form, but long-term over-dosing can be problematic.
Courtesy of Christopher Hawley.

2 gallons of water. Immediately after the contest, she started to suffer ill effects and died within a few hours. These examples are used for some shock value, but each of the chemicals used is very important to human health, if one takes the proper dose. Understanding the science behind the need for these materials and the risk is an important factor.

Acute and Chronic Effects

OSHA 1910.1200 Appendix A provides some criteria for a highly toxic material, which is provided in **TABLE 2-1**. The hazard that a chemical presents is an indicator as to the potential harm it can cause. There are two types of exposures, **acute** and **chronic**, both of which can have serious health effects.

LISTEN UP!

An acute exposure is a quick, one-time exposure to a chemical.

Typically, an acute exposure is one where the body is subjected to a large dose over a short period of time and could have **acute effects**. A chronic exposure is one that occurs over a longer period of time and may be a repeated exposure. Chronic exposures have the potential to cause **chronic effects**. An acute exposure period of time can be from seconds to 72 hours. It should be noted that an acute exposure, depending on the product and

TABLE 2-1 Toxic and Highly Toxic Definitions	
Toxic is a chemical that falls within any of the following categories: ■ A chemical that has a median lethal dose (LD50) of more than 50 mg/kg but not more than 500 mg/kg of body weight when administered orally to albino rats weighing between 200 and 300 g each. ■ A chemical that has a median lethal dose (LD50) of more than 200 mg/kg but not more than 1000 mg/kg of body weight when administered by continuous contact for 24 hours (or less if death occurs within 24 hours) with the bare skin of albino rabbits weighing between 2 and 3 kg each. ■ A chemical that has a median lethal concentration (LD50) in air of more than 200 ppm but not more than 3000 ppm by volume of gas or vapor or more than 2 mg/L but not more than 200 mg/L of mist, fume, or dust when administered by continuous inhalation for 1 hour (or less if death occurs within 1 hour) to albino rats weighing between 200 and 300 g each.	Highly toxic is a chemical that falls within any of the following categories: ■ A chemical that has a median lethal dose (LD50) of 50 mg or less per kilogram of body weight when administered orally to albino rats weighing between 200 and 300 g each. ■ A chemical that has a median lethal dose (LD50) of 200 mg or less per kilogram of body weight when administered by continuous contact for 24 hours (or less if death occurs within 24 hours) with the bare skin of albino rabbits weighing between 2 and 3 kg each ■ A chemical that has a median lethal concentration (LD50) in air of 200 ppm by volume or less of gas or vapor or 2 mg/L or less of mist, fume, or dust when administered by continuous inhalation for 1 hour (or less if death occurs within 1 hour) to albino rats weighing between 200 and 300 g each.

© Jones & Bartlett Learning

its hazards, might cause injuries ranging from minimal to major and potentially leading to death. Overall, the human body does well with short-duration exposures and recovers from them, but all chemical exposures should be avoided. A simple example of an acute exposure is that of a nonsmoker who decides to smoke one cigarette. This one-time exposure for most people is not harmful, nor would it cause any long-term health effects. If an acute exposure causes an immediate effect, this is known as an acute effect.

An example of a chronic exposure, however, is that of the person who smokes three packs of cigarettes a day for 20 years. This person has received doses of cigarette smoke several times a day for a long period of time and is likely to have a chronic effect and health problems associated with this chronic exposure. (Note that abnormalities do exist: The person who smoked one cigarette may develop lung problems from that one acute exposure, and the person who chain smokes for 20 years may never develop health problems associated with the chronic exposure.)

Mercury presents another good example of the differences between acute and chronic, as it is a toxic material, depending on several conditions. An acute exposure to risk to most humans is negligible; women who are pregnant are the exception and are at a higher risk. In some cases, someone could ingest a small amount of mercury and not suffer any ill effects. Where humans are primarily at risk is to a chronic inhalational exposure to mercury. Even very small amounts, like that in a home

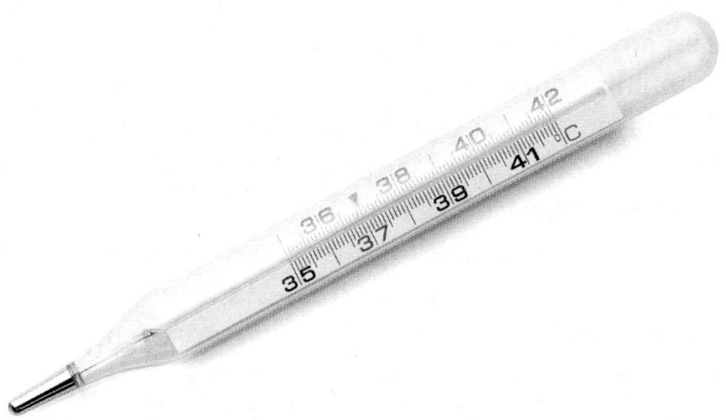

FIGURE 2-2 Mercury found in thermometers, fluorescent lights, and some instruments presents a long-term health risk.
© grey_and/Shutterstock.

medical thermometer, can present a risk over a long period (**FIGURE 2-2**). Care must be taken to properly clean up a spill of this material, as over the long term, mercury vapors can cause damage to a person's health.

Two terms that are related to this discussion are **infectious** and **contagious**; both are used to describe exposures to **etiological** materials. Contagious (communicable disease) means an illness caused by an infectious agent or its toxins that occurs through the direct or indirect transmission of the infectious agent or its products from an infected individual or via an animal, vector, or the inanimate environment to a susceptible animal or human host. Infectious diseases are caused by pathogenic microorganisms such as bacteria, viruses, parasites, or fungi; the diseases can be spread, directly or indirectly, from one person to another. Zoonotic diseases are infectious diseases of animals that can cause disease when transmitted to humans. Secondary contamination can also be contagious for some materials. Some examples include anthrax, which is an infectious pathogen, but it is not contagious through human to human contact. Tularemia is an infectious pathogen that is transmitted through insect bites and is also not transmitted human to human. The flu is infectious and is transmitted human to human via coughing, sneezing, and fluid contact. The time period when someone has been exposed to a pathogen and then begins to show symptoms is known as the incubation period. It is possible that the person is infectious and can spread the infection while not knowing he or she is infectious. The incubation period for the common flu is 1–2 days, and for chickenpox, it is 14–16 days.

SAFETY TIP

Hazardous materials incidents have the ability to impact humans, the environment, and property. In some cases, such as with dioxin, the risk to humans, the environment, and property can last for many years. Through immediate effects or long-term or recurring (chronic) effects, hazardous materials can present a risk. Proper identification of potential hazards and the wearing of proper protective clothing are essential for responder safety.

Types of Hazards

In the realm of hazardous materials, *hazard* is defined as the category of risk that can be inflicted by exposure or contamination with a chemical. There are several methods used to identify possible hazards at a chemical release. The most common method in use today is known by the acronym **TRACEM**, which stands for thermal, radiation, asphyxiation, chemical, etiological, and mechanical hazards. Each of the individual hazards has additional hazards that fit within that classification. Much like the risk-based response theory, the use of TRACEM assigns a chemical to a risk category so that tactical decisions can be based on that classification. The subcategories within TRACEM are:

Thermal

Both heat and cold hazards fit into this category. If a flammable liquid ignites, it is classified as a thermal hazard. If liquefied oxygen contacts your skin, it causes frostbite and a thermal (cold) burn. When liquefied gases are released into small spaces or rooms, the temperature can drop dramatically, presenting hypothermia and cold stress concerns.

Radiation

Types of radiation include nonionizing and ionizing. Nonionizing types include microwave and infrared. Ionizing types such as alpha, beta, gamma, neutron, and X-rays fit into this category. Alpha and beta are particles. Neutron is a form of energy with mass. Gamma and X-rays are forms of electromagnetic energy. Neutron radiation can also transform an item that is not radioactive into one that emits radiation, and it has the capability to irradiate other objects.

Asphyxiation

Both simple and chemical asphyxiants fit into this category. A simple asphyxiant is when inert gases or vapors occupy a space or area and displace oxygen. A chemical asphyxiant is one that prevents the body from using available oxygen.

Chemical

This category includes poisons and corrosives. Poisons may also be referred to as toxic. Some chemicals are highly toxic. There are specific levels of exposure that determine into which category a chemical fits. Corrosives are acid or caustic materials that can damage the skin and/or mucous membranes. Also within this category are **convulsants**, **irritants**, **sensitizers**, and **allergens**. Reactions to a chemical vary person to person much in the way some people are allergic to bee stings while others are not. Chemicals have effects on some people while having no effect on others.

Etiological

Bloodborne pathogens and biological materials are in this category. Etiological agents (infectious substances) are materials known or reasonably expected to contain a pathogen. A pathogen is a microorganism (including bacteria, viruses, rickettsiae, parasites, fungi) or other agent, such as a proteinaceous infectious particle, that can cause disease in humans or animals.

Mechanical

Although not chemical in nature, mechanical hazards exist within the hazard area of a chemical spill. There exists standard slip, trip, and fall hazards that one should always be concerned about. Another example of a mechanical hazard would be getting hit from blast particles, such as from a bomb or boiling liquid expanding vapor explosion (BLEVE). A drum falling on a responder is another example of a mechanical hazard.

Categories of Health Hazards

Within the TRACEM categories, there are terms that responders should understand, because SDSs or industrial contacts may describe some chemicals as fitting into one or more of these categories. One of the most commonly used terms is **carcinogen**, which refers to a material with cancer-causing potential. There are two classifications of carcinogens, known and suspected, with the majority being suspected carcinogens. According to the Chemical Abstracts Service (CAS), a division of the American Chemical Society (ACS), a group that tracks chemicals, there are more than 135 million chemicals in existence today. Each day the ACS adds 4000 more chemicals to their listing. The National Toxicology Program under the Public Health Service of the U.S. Health and Human Services Administration issues an annual report on carcinogens. The 14th annual report lists 62 chemicals that are known to cause cancer and 186 chemicals that are suspected of causing cancer. When dealing with chemical spill response, the risk of getting cancer always causes great fear. Fire fighters are exposed to many chemicals, many of them known cancer-causing agents, but if fire fighters wear their SCBA, these exposures are unlikely to cause problems. For older fire fighters who worked in earlier years when SCBA was not used as extensively as it is today, cancer

is still a leading cause of death, and many retired fire fighters have not had the chance to enjoy retirement due to an early death.

Another term that is commonly used is irritant, which is self-explanatory. An irritant is not corrosive but mimics the effect of a corrosive material in that it can cause irritation of the eyes and possibly the respiratory tract. One notable characteristic of an exposure to irritants is that the effects are easily reversed by exposure to fresh air. Mace, tear gas, and pepper spray are classified as irritants and may be called incapacitating agents by the military.

Sensitizer is a term used to describe a chemical that causes an effect that is an allergic reaction, typically occurring through skin contact. In most cases, an employee can work with a chemical for years and suffer no effects and then one day suffer a severe reaction to the material. Skin reddening, hives, itching, and difficulty breathing are possible symptoms when dealing with a sensitizing agent. Some persons, however, can become sensitive to a chemical after one exposure.

Allergens are chemicals that produce symptoms much like those of sensitizers but are the result of a reaction with an individual's immune system. In many cases, an individual may have been allergic to a substance but was unaware of it until the chemical was encountered. A convulsant chemical is one that can cause convulsions upon exposure, typically by ingestion. There are several drugs that can cause a seizure-like response. **Organophosphate pesticides** make up one group of chemicals that can cause seizure-like activities. The nerve agents grouping of chemical warfare agents are chemically similar to organophosphate pesticides and can cause the same effects. When someone is exposed to an organophosphate pesticide (**OPP**) or a nerve agent, the brain reacts to the chemicals, and seizure-like activity results. There are two symptoms that are classic OPP or nerve agent poisoning: One is pinpointed (small) pupils, and the other is seizure-like activity. Pinpointed pupils can occur with very low doses and in some references is called miosis. Some of the other signs and symptoms are listed in **TABLE 2-2**. One of the other methods used to describe nerve agent or OPP poisoning is through the acronym **SLUDGEM**. The letters stand for salivation (drooling), lacrimation (tearing), urinary, defecation, gastrointestinal upset, emesis (vomiting), and miosis (pinpointed pupils). The bottom line is that if you respond to an incident and a victim is presenting with the loss of a variety of body fluids, having convulsions or tremors, and has pinpointed pupils, he or she has been most likely exposed to an OPP or least likely a nerve agent. In either case, your response and the treatment of the victim are identical.

TABLE 2-2 Common Signs and Symptoms of Nerve Agent and Organophosphate Poisoning

Central Nervous System Signs and Symptoms		
Miosis	Headache	Restlessness
Convulsions	Loss of consciousness	Coma

Respiratory Signs and Symptoms		
Rhinorrhea (severe runny nose)	Bronchorrhea (excessive bronchial secretions)	Wheezing
Dyspnea (shortness of breath)	Chest tightness	Hyperpnea (increased respiratory rate/depth)—early
Bradypnea (decreased respiratory rate/depth)—late		

Cardiovascular Signs Resulting from Blood Loss		
Tachycardia (increased heart rate)—early	Hypertension (high blood pressure)—early	Bradycardia (decreased heart rate)—late
Hypotension (low blood pressure)—late	Arrhythmias (irregular heartbeat)	

Gastrointestinal Signs and Symptoms		
Abdominal pain	Nausea and vomiting	Diarrhea
Urinary incontinence, frequency		

Musculoskeletal Signs and Symptoms		
Weakness (may progress to paralysis)	Fasciculations (muscle twitching)	

Skin and Mucous Membrane Signs and Symptoms		
Profuse sweating	Lacrimation (tears forming)	Conjunctival injection (blood shot eyes)

Adapted from CDC Toxic Syndrome Description, 2005.

In the realm of biological materials two terms are used: Infectious and contagious. Someone who is ill may not present a risk to anyone else. If he or she presents a risk of transmitting his or her illness to another person, then he or she is infectious.

Target Organs

Some chemicals only affect one or more organs and are described as target organ hazards, or they may affect a body system, such as the central nervous system. The effects depend on the individual, dose, concentration, and length of exposure. Some of the target organ descriptions are provided in **TABLE 2-3**.

Routes of Exposure

Exposure is one term that has been used in this chapter that needs further defining, as the mere exposure to a chemical does not mean that there will be an effect. The route of exposure is one major factor in determining the hazard that a chemical may present to humans. One other factor is the issue of each human being slightly different. As with people who are allergic to bee stings or

TABLE 2-3 Target Organs and Systems

Name	Target Organ or System
Central nervous system (CNS) chemicals	Affect the central nervous system and can cause short-term or long-term effects. Commonly short-term memory is lost after exposure to a CNS hazard material. Many of the hydrocarbons cause CNS effects, and people sometimes purposely expose themselves to a CNS agent to receive a "high" from the exposure (huffing). In the long term, the brain cells are damaged, never to recover. Neurotoxins essentially cause the same effects.
Peripheral nervous system (PNS) chemicals	Much like CNS chemicals, the PNS chemicals affect the body's ability to move in a coordinated fashion. Exposure to a PNS chemical causes a disruption of the brain's ability to move messages to the other body systems.
Hepatotoxins	These types of materials affect the liver and, if the exposure is high enough, can cause severe damage to the liver.
Nephrotoxins	These materials adversely affect the kidneys.
Reproductive toxins	These toxins affect the ability to reproduce and can cause birth defects. These types of toxins can stay within the body, so they can have adverse effects on a pregnancy even if the exposure occurred a considerable time before the pregnancy.
Mutagens	An exposure to a mutagen may not cause any harm to the people who received the exposure, but the effect might be transmitted to their offspring. Mutagens might cause damage to the genetic system and can cause mutations that can become hereditary.
Teratogens	These materials can affect an unborn child (embryo and/or fetus) causing birth defects.

© Jones & Bartlett Learning

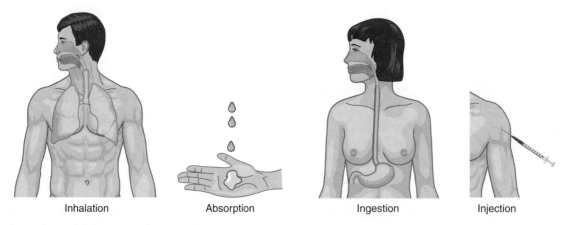

| Inhalation | Absorption | Ingestion | Injection |

FIGURE 2-3 Respiratory system route of exposure.
© Jones & Bartlett Learning

food, color blind, or left-handed, each person's chemistry is slightly different. What may have a major impact on one person may not have any impact on another.

The four primary routes of exposure are inhalation, absorption, ingestion, and injection (**FIGURE 2-3**). The route that is the most commonly associated with causing health effects, both acute and chronic, is the respiratory system route.

SAFETY TIP

In almost all cases, the respiratory system requires some type of protection. For emergency services workers, this is the easiest system to protect because they have easy access to SCBA, should be familiar with it, and are comfortable with its use.

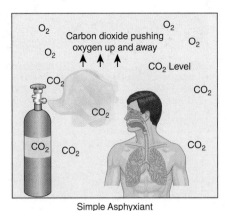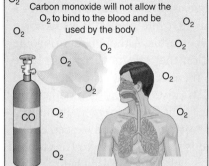

Simple Asphyxiant   Chemical Asphyxiant

FIGURE 2-4 Examples of simple and chemical asphyxiants.
© Jones & Bartlett Learning

The respiratory system can be affected by gases, vapors, and solid materials such as dust and other particles. In many cases, the chemicals may not have any effect on the respiratory system itself but may enter the body through the respiratory system and affect other organs or body systems. When dealing with respiratory hazards, there are two categories of asphyxiants: Simple and chemical (**FIGURE 2-4**). Although the result is usually the same, the way a person is killed does differ significantly. Simple asphyxiants displace the oxygen in the air and push it out of the area. It is not an adverse chemical reaction but simply a matter of something other than oxygen occupying the space in the body where the oxygen should be. Normal oxygen levels are 20.9 percent in air, and the body starts to develop difficulty breathing at less than 19 percent. A person starts to have serious problems at less than 16.5 percent oxygen. **TABLE 2-4** lists the impact of reduced oxygen to humans. Gases such as nitrogen and halons move oxygen out of the area and cause people in the area to have difficulty breathing. If the concentrations are high enough, death could result. In Leon, Mexico, in 2013 at a Jägermeister™ sponsored pool party, several people emptied nitrogen dewars into the pool. Initially, everyone thought the developing cloud was interesting, until people started to pass out. No one died, but eight people became oxygen deficient and passed out. Initial media reports mistakenly stated it was a reaction with the pool chemicals, when it was just the change of the state of matter for the liquid nitrogen. Chemical asphyxiants work in a different manner. They cause a chemical reaction within the body and do not allow it to use the readily available oxygen. The most common chemical asphyxiant is carbon monoxide (CO). When CO is in the air in sufficient quantities, it enters the bloodstream through the lungs. It binds with the hemoglobin in the blood, forming carboxyhemoglobin. Because hemoglobin has an attraction for CO about 225

TABLE 2-4 Oxygen Deficiency Impact	
Oxygen Level	**Impact**
19%	Minor psychological impact, person may not be aware of the impact.
15–19%	Impaired thinking, reduced physical capabilities, increased pulse and respiratory rate. Person may not be totally aware of the impairment.
12–15%	Significant impairment to judgment and physical ability.
10–12%	Extreme difficulty in respirations, significant physical impact, with nausea and vomiting. Victims may have blue lips.
8–10%	Unconsciousness, vomiting, and blue lips within minutes.
6–8%	Quick loss of consciousness, without rescue in 6–8 minutes death likely.
<6%	Immediate loss of consciousness, convulsions, and imminent death. Brain damage most likely to occur even if victim survives.

Excerpted from a variety of gas supply safety documents.

times greater than that of oxygen, it does not allow the oxygen molecules to bind with the blood, which causes severe health problems and often death.

Another common chemical asphyxiant is cyanide. Cyanide exposure in the body disrupts cellular respiration, which prevents the cells from using oxygen no matter how much oxygen is present. This is of concern to the fire service, as various forms of cyanide are often found as a product of combustion in structure fires involving modern construction materials.

Although it can be confusing, carbon dioxide is often thought of as a simple asphyxiant, but it is more of a chemical asphyxiant. CO_2 intoxication is also known as hypercapnia or hypercarbia. Although there is some aspect of asphyxiation that affects persons in large CO_2 environments, there is a chemical aspect within the body as well. CO_2 exists within the bloodstream as it is the byproduct of breathing oxygen. In normal circumstances, one breathes in air, and oxygen enters the bloodstream and is used by the lungs. The carbon dioxide dissolves in the blood or binds to hemoglobin at the tissue level, travels through the bloodstream to the lungs, and in the alveoli the CO_2 diffuses off the hemoglobin/out of the blood and into the airspace of the lungs due to the concentration difference/pressure differential between the bloodstream and the air in the lungs (CO_2 partial pressure). The same process, just in reverse, causes oxygen to bind to the hemoglobin (oxyhemoglobin). CO_2 then is exhaled, and the process repeats itself. With a higher amount of CO_2 in the air, the pressure differential is reduced or reversed, preventing CO_2 from diffusing out of the bloodstream, or at high ambient CO_2 concentrations even causes more CO_2 to enter the bloodstream in the lungs, creating respiratory issues. The CO_2 also alters the chemistry of the blood, making it acidic in nature, causing acidosis. When blood chemistry is altered, one can suffer from a multitude of cardiac problems, namely arrhythmias. In addition, high levels can cause headaches, dizziness, and altered mental status and the inability to carry out normal physical tasks. **TABLE 2-5** provides a listing of the effects from CO_2 exposure. There are also issues with low-dose exposures to CO_2 over the long term. Exposures over a period of a few days to weeks can cause altered mental status, lung volume concerns, altered blood chemistry, and cardiac concerns. Chapter 4, Oxygen Level Determination, provides several case studies regarding CO_2 exposures, which are going to increase.

The other common route of entry is via skin absorption, as the skin is the body's largest organ. Although some chemicals can cause damage to the skin and may irritate it, this does not mean that it is toxic by skin absorption. The number of chemicals that are toxic by skin absorption is relatively low, but precautions should be taken to minimize contact and, if possible, have no skin contact with chemicals. Skin contact with chemicals can cause burns, rashes, or drying of the skin. The only way to provide skin protection is to wear proper protective clothing that does not allow the chemicals to get onto the skin. Fire fighter turnout gear slows the process down but does not prevent the eventual migration of the chemical to the skin. Effective decontamination is required to ensure that all the chemical is cleaned from the skin.

TABLE 2-5 CO_2 Exposure Symptoms	
Amount	**Symptoms**
10,000–15,000 ppm (1–1.5%)	Affects blood chemistry after a few hours.
30,000 ppm (3%)	Alters the mental state, increased respiratory and pulse rate, high blood pressure, headaches, and impacted hearing.
40,000–50,000 ppm (4–5%)	Rapid breathing, altered state after 30 minutes.
50,000–100,000 ppm (5–10%)	Difficulty breathing, headache, and altered mental status.
100,000–1,000,000 ppm (10–100%)	Levels about 100,000 ppm will cause unconsciousness is minutes, and higher levels may be fatal.

Excerpted from European Industrial Gas Association Safety Advisory Council Bulletin.

The other route of entry is ingestion, which is more common than one would think. The most common method of chemical ingestion is from improper decontamination techniques. It is also possible that food and drinks served during rehabilitation could become contaminated depending on their location regarding the incident.

One other route of exposure is injection, although it is not considered to be one of the major routes. Many emergency services workers are exposed to hazardous materials via this route, and for this population, it is probably one of the leading routes, after inhalation. The most common material that emergency services workers are exposed to is body fluids or what is referred to as bloodborne pathogens. Other methods of injection from being near a high-pressure line when it breaks, such as a hydraulic rescue tool fluid line that would inject hydraulic fluid into a person's body. Other than standard infection control practices, there is little protection for these types of exposures except to properly wear full PPE when working in and around situations where exposure to these fluids is possible.

Factors that affect the rate of exposure, regardless of the route, are basic items such as temperature, pulse, and respiratory rate. The higher each of these items is in an individual, the more likely it is that the chemical will have some effect. The damage that chemicals have is based on the equation $Effect = Dose \times Concentration \times Time$.

SKIN HAZARDS AND CHEMICAL EXPOSURE

In 1988, NIOSH had adopted skin hazard notations for 142 chemicals listed in the NIOSH Pocket Guide for Chemical Hazards. The problem was that the skin notation did not specify the hazard to the skin or whether the chemical was toxic by skin absorption or was a corrosive hazard. At the time, there was little science or standardization to the notations. In 2009, NIOSH published a new strategy for assigning new NIOSH skin notations. In the system, skin is abbreviated to SK and identified several methods of skin notations. The new notations are systemic (SYS), direct (DIR), and sensitizing (SEN). Chemicals that are highly toxic and that can cause fatal effects are noted with the subnotation (FATAL). Two other subnotations used are irritant (IRR) and corrosive (COR). The system identifies a skin hazard for a chemical that can cause irritation by direct contact by the notation SK:DIR (IRR). Other abbreviations used include SK, which indicates that the reviewed data showed no hazard with the skin. If there is insufficient data with skin hazards, ID$^{(SK)}$ is used. When a chemical has not been evaluated, ND is used.

For the chemicals with a skin notation, NIOSH used phenol as an example. Phenol is also known as carbolic acid, benzenol, phenolic acid, and several other synonyms. The skin notation for phenol is SK: SYS-DIR (COR). In analysis of literature, NIOSH found that phenol affects the central nervous system and causes respiratory depression, cardiac arrest, body weight changes, and decreased survival from direct skin contact. They also found sufficient evidence that it was corrosive to the skin. The science behind the notation showed that when applied to the skin the absorbed dose was 4–23 percent of the amount applied, but the amount varied with the concentration and time applied. One study showed 12.6 percent of the dose applied was absorbed in 30 minutes. At 60 minutes, 22.7 percent was absorbed. In test subjects with respiratory protection, whole body exposures showed that phenol moved through the skin with a 6-hour exposure. In animal studies, the cumulative skin absorption was 66–80 percent after 120 hours. Studies have also shown that skin contact causes reddening skin, inflammation, discoloration, eczema, severe edema, necrosis, and chemical burns.

Modified from Current Intelligence Bulletin 61 A Strategy for Assigning New NIOSH Skin Notations.

This equation relates to acute and chronic exposures. A small dose, that is, a small concentration for a short time, is not likely to have an adverse effect on a normal human. A large dose at a high concentration over a long period of time will in most cases have an effect. People exhibiting normal vital signs who are exposed to a chemical may not have any effects, but if they jog around the block prior to being exposed to a material, they may be affected. An increased pulse rate allows for chemicals to spread throughout the body faster. In some confined space incidents where chemicals may have played a factor in injuries and deaths, increased vital signs play a role. In most cases, the first victim may be unconscious. The body's system has slowed down, and in some cases, the victim went down due to a lack of oxygen. When a person recognizes that a coworker has gone down, the vital signs increase, and when trying to perform a rescue, the person may be exposed to a higher level of the chemical than the coworker he is trying to rescue. This is one of the reasons why, in some cases, the rescuer dies, and the original victim ends up surviving.

LISTEN UP!

An increased temperature of the body allows for faster absorption into the body, and the accompanying increased respiratory rate causes more chemicals to enter the respiratory system.

Exposure Levels

In industry, monitoring for exposures is commonplace and is usually a preventive action, but in the emergency services, it can be an afterthought, typically after an incident has occurred. Several different types of exposure values have been issued by a variety of agencies, some of which can be very confusing. Exposure values have been established for the commonly used chemicals and for a variety of situations. The key to preventing exposures is to monitor for hazardous materials and to wear appropriate PPE. The one key agency involved with exposure values is the Occupational Safety and Health Administration (OSHA). The exposure values that they set are the ones that must be followed by all industries, including the fire service because it too is considered an industry. Another organization that issues exposure values is the American Conference of Governmental Industrial Hygienists (ACGIH), a group that advocates worker safety and conducts studies regarding chemical exposures. The National Institute of Occupational Safety and Health (NIOSH), a research arm of OSHA, issues recommendations for exposure levels. OSHA is the only agency that provides legally binding exposure values; all others are recommendations. In some OSHA regulations, the employer is required to use the lowest published exposure values, which in many cases are not OSHA's own values. When dealing with emergency situations, it is always recommended that responders follow the lowest published values.

The exposure values have widely varied criteria, but some are based on an average male and are for an

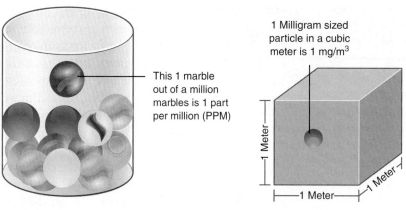

FIGURE 2-5 Explanation of units of measure.
© Jones & Bartlett Learning

industrial application. The permissible exposure limit (PEL) is based on an 8-hour day, 40-hour workweek with a 16-hour break between exposures. The values that are issued for the various substances are typically conservative; in a given population, it is not atypical to find someone who is sensitive to a chemical at a lower value than the rest of the group. Each value has an extra margin for safety built in, up to 10,000 times the actual value. The exposure values are typically listed in parts per million (ppm) or as milligrams per meter cubed (mg/m^3). Explanations of these terms are provided in **FIGURE 2-5**.

The most common exposure values are expressed in ppm, but for some materials (typically solids), the values may be expressed as mg/m^3. In some cases, the values may be listed in parts per billion (ppb), which are very small amounts. The values are generally for 8 hours. OSHA refers to this 8-hour exposure value as the **permissible exposure limit (PEL)**. The ACGIH refers to this 8-hour exposure as the **threshold limit value (TLV)**. Both are average exposures over the 8-hour period. A worker can be exposed to more than the PEL or TLV if at the end of the day the exposure value is less than the PEL or TLV. In some cases, these exposure values are called threshold limit value time weighted averages (TLV-TWAs), and they may be expressed as the OSHA-TWA or the ACGIH-TWA, which is an 8-hour daily average exposure. NIOSH issues **recommended exposure limits (RELs)**, which are for a 10-hour day, 40-hour workweek. These exposure values are outlined in graphic form in **FIGURE 2-6**.

Other values that may be listed for chemicals are the **ceiling levels**, generally referred to as PEL-C or TLV-C. These provide an amount that is the highest level to which an employee can be exposed. When figuring an average, there are times when employees are going to be exposed to chemicals at a level higher than the PEL or TLV, but there are also times when

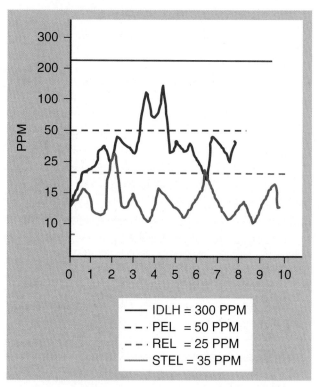

FIGURE 2-6 A graphic example of exposure levels, showing the IDLH, which requires immediate withdrawal from the area. The PEL is an average exposure that has to be less than 50 ppm over an 8-hour period. The REL is an average exposure of a 10-hour period. The STEL is an average of a 15-minute exposure.
© Jones & Bartlett Learning

they will be exposed to less than those levels. If their exposure average is less than the PEL or TLV, they are acceptable. Certain chemicals can cause effects at levels that may be obtained through worker exposure and would not be considered safe, but the overall exposure would fall below the PEL or TLV. To avoid unsafe levels, the safety organizations may attach a ceiling level to an exposure value, and that is the highest level that the employee can be exposed to, regardless of what the end average is.

Another exposure value is known as the threshold limit value **short-term exposure limit (TLV-STEL)**. This value is assigned to a 15-minute exposure. An employee can be exposed at this level for 15 minutes and then is required to take an hour break from the exposure. The employee can do this four times a day without any adverse effects. NIOSH is also using an excursion value that is coupled with a time limit, 5–30 minutes typically. At this level, an employee can enter an environment one time and not suffer any effects.

The last value can be confusing because it is called the **immediately dangerous to life or health (IDLH)** value. One would think that being exposed to a chemical at the IDLH level would mean that death might be imminent. This is a value that is the maximum airborne concentration that an individual could escape and not suffer any adverse effects. The actual definition does not match the legal definition. At the IDLH level, emergency responders need to be using SCBA as an absolute minimum, and if the chemical is toxic by skin absorption, then a fully encapsulated vapor tight suit (Level A) must be used.

Other values that a responder may see are called **lethal doses (LD$_{50}$)** or **lethal concentrations (LC$_{50}$)** . The LD is for solids and liquids, and the LC is for gases. In most cases, animal studies provide these values, but some are derived from human studies, suicides, and murders. The 50 attached to the LD or LC means 50 percent of the exposed population. In the studies, a certain number of test subjects were exposed to a low level of chemicals. After a period of time, the test subjects were studied for any adverse effects. Another group was exposed to a higher level of the chemical.

When the subjects were exposed to a certain level and 50 percent of the test subjects died, this established the LD$_{50}$ or LC$_{50}$. With terrorism, emergency responders use military data, much of which is based on LD$_{50}$ or LC$_{50}$ type studies. The military values are generally expressed as LCt$_{50}$ or lethal concentration to 50 percent of the population, with "t" representing time, usually expressed in minutes. The military also uses ICt$_{50}$, which is the **incapacitating concentration** to 50 percent of the population in a certain amount of time. **TABLE 2-6** lists exposure values for some chemicals. Two other terms that are useful when discussing exposure values are threshold of detection and odor threshold. When discussing detection devices, it is important to know what is the lowest level that the device can detect, which is threshold of detection. There are some devices that have thresholds of detection above the IDLH value for some materials, which means the device is not very effective at providing protection. Odor threshold is a bit more complicated as it relates to the human's ability to smell a material. In some cases, the odor threshold is expressed as an average or is expressed with a percentage affixed; that is, 50 percent of the population can smell this chemical. Hydrogen cyanide is an example in which the "characteristic" burnt almond odor can only be detected by 60–80 percent of the population. For those who can smell the odor, they may suffer from olfactory fatigue, and the longer they are exposed, the less likely they are to smell the odor. One form of mercaptan, tertiary butyl mercaptan, is a common odorant added to natural gas (methane). All mercaptans are irritating, and after time, humans become desensitized to them and may not smell the odorant in the gas, which can have disastrous consequences.

TABLE 2-6 Exposure Values						
Chemical	**Vapor Pressure (mmHg @ 25°C/77°F)**	**Odor Threshold (ppm)**	**PEL (ppm)**	**IDLH (ppm)**	**LCt$_{50}$ (ppm)**	**ICt$_{50}$ (ppm)**
Ammonia	7500	0.4	50	300	1107	77
Chlorine	5830	0.002	1	30	6551	620
Hydrogen cyanide	742	0.58	10	45	3600	n/a
Mustard Agent	0.11	<1.3 mg/m^3	0.0005	0.0005	231	30
Phosgene	1420	0.4	0.1	2	791	395
Sarin	2.86	1.5 mg/m^3	0.000017	0.03	12	8

HYDROGEN CYANIDE EXPOSURE LEVELS

Hydrogen cyanide is a common toxic material that is found in several locations. It is common in the plating industry and was formerly used as a fumigant. It is common in the plastics industry. The exposure values for hydrogen cyanide with a skin notation are:

PEL: 10 ppm   REL: 4.7 ppm   STEL: 4.7 ppm   IDLH: 50 ppm

In OSHA's study of hydrogen cyanide (HCN), they provide a series of time variables, which include:

Immediately fatal: 270 ppm

Fatal after 10 minutes: 181 ppm

Fatal after 30 minutes: 135 ppm

Fatal after 0.5–1 hour or later (dangerous to life): 110–135 ppm

Tolerated for 0.5–1 hour without immediate or late effects: 45–54 ppm

Slight symptoms after several hours: 18–36 ppm

There is limited information on the skin absorption of cyanide, and much of the information is 60 or more years old. One study showed that three men wearing SCBA were in atmosphere of 20,000 ppm for 8–10 minutes. They suffered from dizziness, weakness, and throbbing pulse that lasted a few hours, but all made a full recovery. In animal studies, toxicity was not observed after an exposure of 4975 ppm for 3 hours. Exposures at 13,400 ppm for 47 minutes did result in some deaths, which indicated skin absorption.

Acute Exposure Guidelines (AEGL)

One other value you will see in a variety of reference materials is the Acute Exposure Guidelines (AEGL), which are issued by a committee coordinated by the EPA. These exposure values are primarily for planning purposes, but they do have some applicability for emergency response work. They are developed for a release of 10 minutes up to 8 hours. There are three levels provided, and they are defined as:

AEGL-1 is the airborne concentration, expressed as parts per million or milligrams per cubic meter (ppm or mg/m^3) of a substance above which it is predicted that the general population, including susceptible individuals, could experience notable discomfort, irritation, or certain asymptomatic nonsensory effects.

However, the effects are not disabling and are transient and reversible upon cessation of exposure.

AEGL-2 is the airborne concentration (expressed as ppm or mg/m^3) of a substance above which it is predicted that the general population, including susceptible individuals, could experience irreversible or other serious, long-lasting adverse health effects or an impaired ability to escape.

AEGL-3 is the airborne concentration (expressed as ppm or mg/m^3) of a substance above which it is predicted that the general population, including susceptible individuals, could experience life-threatening health effects or death.

Levels below the AEGL-1 may still cause some impact but are not likely to cause harm. Some examples include:

TABLE 2-7 Acute Exposure Guidelines

Ammonia

	10 Mins	30 Mins	1 Hour	4 Hours	8 Hours
AEGL-1	30 ppm	30 ppm	30 ppm	30 ppm	30 ppm
AEGL-2	220 ppm	220 ppm	160 ppm	110 ppm	110 ppm
AEGL-3	2700 ppm	1600 ppm	1100 ppm	550 ppm	390 ppm

Chlorine

	10 Mins	30 Mins	1 Hour	4 Hours	8 Hours
AEGL-1	0.5 ppm	0.5 ppm	0.5 ppm	0.5 ppm	0.5 ppm
AEGL-2	2.8 ppm	2.8 ppm	2 ppm	1 ppm	0.71 ppm
AEGL-3	50 ppm	28 ppm	20 ppm	10 ppm	7.1 ppm

TABLE 2-7 Acute Exposure Guidelines (*continued*)

Hydrazine

	10 Mins	30 Mins	1 Hour	4 Hours	8 Hours
AEGL-1	0.1 ppm	0.1 ppm	0.1 ppm	0.1 ppm	0.1 ppm
AEGL-2	23 ppm	16 ppm	13 ppm	3.1 ppm	1.6 ppm
AEGL-3	64 ppm	45 ppm	35 ppm	8.9 ppm	4.4 ppm

© Jones & Bartlett Learning

After-Action REVIEW

IN SUMMARY

Although toxicology and exposure values can be confusing, it is important to know what each of the values means and how to apply their use in an emergency response. At 9 PM a 911 call is received for smoke coming from an industrial location where there are several facilities. A first-alarm fire assignment is dispatched. Another caller reports that the facility is at a fruit packing company and provides a specific address. Once the 911 center identifies the specific address, they note that the facility stores anhydrous ammonia. In many jurisdictions, a hazardous materials team would be dispatched or at least alerted at this point. This information was provided to the fire department through Tier 2 reports, and as it is an extremely hazardous substance (EHS), they also supplied SDSs to the fire department.

Additional calls to the 911 center are reporting an ammonia odor in the adjacent neighborhood. The first arriving fire company notices that there is a white vapor cloud coming from the structure. The white vapor cloud is ammonia, and there is a chemical release and not a fire. Anhydrous ammonia is a toxic, caustic, and flammable gas, which can travel long distances. It is a lighter than air gas and, in most cases, rises, but in some weather conditions, it stays low. In either case,

air monitoring needs to be done in the downwind neighborhoods. Responders also need to know the health hazards of the material, the signs and symptoms of exposure, routes of entry, and the various exposure values. All this information can be obtained from a variety of sources, which include information filed with the Tier 2 reports. It can also be obtained from SDSs, which should be at the facility or can be obtained from CHEMTREC/CANUTEC/SETIQ, who can interface with the manufacturer or the shipper. CHEMTREC/CANUTEC/SETIQ can also provide information on the material in addition to the SDS.

The information for anhydrous ammonia includes:
It is a caustic liquefied gas that has a vapor density of 0.59 and a vapor pressure of 7510 mmHg @ 77°F (25°C) and a pH of 11. It has a flammable range of 16–25 percent. IDLH is 300 ppm, REL is 25 ppm, PEL is 50 ppm, and it has a STEL of 35 ppm. This information will help the incident commander determine the isolation areas, proper protective clothing, decontamination required, and health effects. Air monitoring will establish the levels of ammonia and determine the areas of risk. In future chapters, we will discuss some additional information on how to use this information.

KEY TERMS

Acute A quick, one-time exposure to chemicals.

Acute effect An acute exposure causes an immediate effect.

Allergen Material that causes a reaction to the immune system.

Carcinogen Material capable of causing cancer in humans.

Ceiling level Highest exposure a person can receive without suffering any ill effects; combined with the PEL, TLV, or REL, it establishes a maximum exposure.

Chronic Continual or repeated exposure to a hazardous material over time.

Chronic effect Repeated exposure several times a day for a long period of time.

Convulsant A chemical that can cause seizure-like activity.

Etiological Includes biological agents, viruses, and other disease-causing materials.

Highly toxic Toxic material that can cause harm in very small doses.

Incapacitating concentration (ICt_{50}) Military term for incapacitating level set over time to 50 percent of the exposed population.

Industrial hygienist A person trained in occupational health and safety.

Infectious Caused by pathogenic microorganisms, such as bacteria, viruses, parasites, or fungi; the diseases can be spread, directly or indirectly, from one person to another.

International units (IU) Unit of measure for pharmacists based on the effect on the human body.

Irritant Material that is irritating to humans but usually does not cause any long-term effects.

Lethal concentrations (LC_{50}) Term used to describe the amount of vapors or gas that was inhaled to cause harm or death to the exposed test animals. A lethal concentration of 50 percent means that the dose killed half the animals that were exposed to that level.

Lethal dose (LD_{50}) Term used to describe a dose of a solid or a liquid that caused harm or death to test animals. A lethal dose of 50 percent means that the dose provided killed half the animals.

Organophosphate pesticide (OPP) Toxic group of chemicals that can cause seizure-like activities. Nerve agents are chemically similar to OPP and can cause the same effects.

Permissible exposure limit (PEL) OSHA value that regulates the amount of chemical that a person can be exposed to during an 8-hour day.

Recommended exposure level (REL) Exposure value established by NIOSH for a 10-hour day, 40-hour workweek. Is similar to PEL and TLV.

Safety Data Sheet (SDS) A form with chemical information that lists properties, health effects, and emergency actions.

Sensitizer Chemical that after repeated exposure may cause an allergic effect for some people.

Short-term exposure limit (STEL) Fifteen-minute exposure to a chemical; requires a 1-hour break between exposures and is only allowed four times a day.

SLUDGEM Acronym for salivation (drooling), lacrimation (tearing), urinary, defecation, gastrointestinal upset, emesis (vomiting), and miosis (pinpointed pupils).

Threshold limit value (TLV) Exposure value that is similar to the PEL but issued by the ACGIH. It is for an 8-hour day.

TRACEM Acronym for the types of hazards that may exist at a chemical incident; thermal, radiation, asphyxiation, chemical, etiological, and mechanical.

Toxicology The study of poisons.

Review Questions

1. What is the best reference source for chemical information?

2. What is the best online reference source for chemical information?

3. What are the four routes of exposure?

4. What exposure level indicates immediate danger?

5. What exposure level is measured in 15-minute increments?

6. A long-term (over years) exposure is known as?

CHAPTER 3

Identifying the Corrosive Risk

OUTLINE

- Learning Objectives
- Introduction
- Acids
- Bases
- Methods of pH Detection
- In Summary
- Key Terms
- Review Questions

LEARNING OBJECTIVES

Upon completion of this chapter, you should be able to:

- Identify the importance of identifying a potential corrosive risk
- Describe the common methods of identifying the corrosive risk
- Describe the working definition of a corrosive material

Hazardous Materials Alarm

Hazardous Materials Alarm - Contagious Asthma Patients—While headed to lunch one day, my boss and I heard a dispatch for two kids having an asthma attack. Being inquisitive and wondering why two kids would be having simultaneous asthma attacks at an elementary school, he called dispatch for more information. Dispatch did not have much additional information other than the kids were in a classroom and started to have an asthma attack, and now a third was also having trouble. Since the cause was probably chemical related, a hazardous materials assignment was requested. When crews arrived, they found the kids had been evacuated into the schoolyard. The teacher of the classroom in question was asked about possible causes. After much questioning, it was determined that the students started having problems after the teacher sprayed a cleaning solution on the students' tables. She did not know what kind it was but could show us the bottle, which was a household cleaning solution. We asked if the label matched the contents, and she said that the janitor would fill it as needed. The janitor said he filled the bottle from a 55-gallon drum in the shop. He did not have a safety data sheet (SDS) for the solution, but the drum was labeled. We asked if he diluted the cleaner in the spray bottle, to which the answer was no. In the drum was pure potassium hydroxide, a very serious corrosive with a pH of 13, which obviously and understandably irritated the students.

1. What was the first clue that this was an airborne issue?

2. What if an SDS was not available?

Introduction

Because humans and electronic devices usually are harmed when exposed to corrosive materials, determining the pH of a released material is very important. The reasons are obvious: **Corrosive** materials burn or irritate humans and quickly render any electronic instrument useless. Most electronic detection devices may work for a limited time in a corrosive atmosphere but eventually stop working, sometimes unknown to the user. pH paper is one of the most useful tools that a responder can depend on to identify a corrosive risk. It is recommended that pH paper be one of the first items available at a chemical release, and the easiest way to do that is to attach a strip of pH paper to the other instruments. Also, determining the pH of a hydrolysis material is one of the methods of detecting warfare agents.

The term "corrosivity" is a term that is applied to both **acids** and **bases** and is used to describe a material that has the potential to corrode or eat away skin or metal, and some examples are shown in **TABLE 3-1**. We deal with corrosives every day: Our stomach is naturally acidic, and our heart relies on an acid–base relationship, and we use many corrosive materials in our everyday lives. Acids are sometimes referred to as corrosives, while bases are also known as **alkalis** or **caustics**. The accurate way to describe a corrosive is to identify the material's pH, which provides some measure of corrosiveness. **pH** is an abbreviation for several terms: potential of hydrogen, power of hydrogen, percentage of hydrogen ions, and the negative logarithm of hydrogen ion activity. It is used to designate the corrosive nature of a material. Acids have hydronium ions (H^+), and bases have hydroxide ions (OH^-). It is the concentration of hydrogen ions that makes up the pH number. If a substance has more hydronium ions than hydroxide ions, it is acidic. With weak acids such as acetic acid, the term "dissociation" is discussed, and it relates to the ability of a hydrogen ion to come off a chemical, in other words changing its electrical charge. Acids in water dissociate to varying degrees and release hydrogen ions in solution. Strong acids have the potential to react with metals and selected other materials and produce hydrogen gas, which can also

TABLE 3-1 Chemical and Physical Properties of Common Corrosives

Name	pH	Vapor Pressure	Freezing Point	Notes
Acetic acid (vinegar)	3	11 mmHg	62°F (17°C)	Colorless liquid
Acetone*	4–5	231 mmHg @ 77°F (25°C)	−95°F (−71°C)	Colorless liquid
Anhydrous ammonia	13	8.5 atm	−108°F (−77°C)	Colorless gas
Aqueous ammonia	11–12			
Ethyl alcohol	4	59 mmHg @ 77°F (25°C)	−114°F (−81°C)	Colorless liquid
Gasoline*	4–5	> 300 mmHg @ 100°F (38°C)	~ −130°F (~ −90°C)	Colorless to slightly yellow or brownish liquid
Hydrochloric acid −38%	0–1	212 mmHg	−173°F (−114°C)	Colorless to slightly yellow
Hydrofluoric acid	0	917 mmHg @ 77°F (25°C)	−117°F (−83°C)	Colorless
Nitric acid	0	48 mmHg	−44°F (−42°C)	Colorless to slightly yellow liquid. Yellow or red liquids are usually fuming nitric acid.
Oleum − 50% sulfur trioxide	0	159 mmHg @ 104°F (40°C) 50% SO$_3$		Colorless to slightly brown
Orange juice	3		~ 32°F (0°C)	Orange
Pepsi™	2–3		~ 32°F (0°C)	Caramel
Phosphoric acid	1	0.03 mmHg		Crystalline solid but usually found mixed with water; colorless
Sodium hydroxide (lye)	13			Crystalline solid but may be found mixed with water; Colorless to off white
Sulfuric acid	0	0.001 mmHg	51°F (10°C)	Colorless to brown liquid

*Materials such as hydrocarbons are slightly corrosive but are not classified as such as they do not contain hydrogen ions or hydroxide ions, so they do not indicate accurately on pH paper. A complication is that when pH paper is dipped into one of these materials the wet paper appears to provide a pH reading, but in reality, it does not. Experienced responders recognize the leading edge of the color change and report an accurate pH.
© Jones & Bartlett Learning

present a fire hazard. Materials having a pH of less than 7 are considered acids, and materials with a pH greater than 7 are considered bases. A material having a pH of 7 is considered neutral, having equal concentrations of hydronium and hydroxyl ions. In legal terms (and by no means chemically), in most states a pH between 5 and 9 is legally neutral and can be disposed of without any permits or special conditions.

pH indicator papers are paper or strips that indicate pH levels, and there are several different types. They all operate on the same principle, as they are impregnated with one or more pH reacting dyes, usually an acid or a base. When the dye encounters a corrosive, it changes color. Some pH paper or strips are for narrow detection, such as a range of 1–1.5, and have test spots for 1, 1.1, 1.2, 1.3, and 1.5. For emergency response work, a multirange paper or strip works best. The most common pH testing paper has a scale of 0–13 but is available in a variety of pH ranges (**FIGURE 3-1**). This multirange paper is impregnated with a variety of pH reacting dyes, which indicate the pH level. For emergency response work, identifying a pH to a whole number is more than adequate. Knowing that a material has a pH of 1 is adequate; a responder does not need to know that a material has a pH of 1.2 to take appropriate action.

The pH scale is a logarithmic scale, meaning the movement from 0 to 1 is an increase by a factor of 10. The movement from 0 to 2 is an increase of a factor of 100. A material with a pH of 2 is 10 times less acidic than one with a pH of 1. A material with a pH of 3 is 100 times more acidic than one with a pH of 5. Some responders forget about this logarithmic scale when neutralizing a corrosive material at a spill and keep adding large quantities of a base material to a spilled acid. It takes a lot of a weak base to move an acid from a pH of 1 to a pH of 2, and, as the pH gets closer to a neutral value of 7, it takes a lot less of the base material. In some cases, too much base is added; then responders must add a weak acid to the spill to bring the material back to neutral.

LISTEN UP!

Most electronic detection devices may work for a limited time in a corrosive atmosphere but eventually stop working.

Another factor that must be considered when dealing with corrosives is concentration. Most corrosives have some amount of water in them, and some are diluted to low percentages. As an example, sulfuric acid comes in a range of concentrations from 2 percent (spent pickle liquor), to the normal range of 30–37 percent for battery electrolyte, to 98 percent lab and commercial concentration. One unusual form of sulfuric acid is

oleum, which is concentrated sulfuric acid that exists as 120–160 percent sulfuric acid. Oleum is supersaturated with sulfur trioxide. Normally, sulfuric acid does not have much vapor pressure (1 mmHg at 294°F [146°C] or 0.001 mmHg at 68°F [20°C]), but oleum, also called fuming sulfuric acid, has a high vapor pressure due to the added sulfur trioxide. When it is released, it interacts with water vapor in the air and creates a significant vapor cloud of concentrated sulfuric acid. It is very reactive and produces a significant corrosive vapor cloud. The more moisture in the air, the worse the cloud will be.

SAFETY TIP

Chemicals with a pH of less than 2 or more than 11 present a significant risk for injury to humans.

Acids

The most common acids are sulfuric, hydrochloric, nitric, phosphoric, and hydrofluoric, all of which have a pH between 0 and 1. Phosphoric acid is used for a number of purposes, but it is most commonly encountered in soft drinks such as Pepsi™ and Coke™. One would think that having an acid in a common drink would be an issue, but there are several reasons that one shouldn't worry. Primarily the estimated concentration is less than 1 percent. The pH of phosphoric is not much different from the citric acid found in orange juice, and it's the acidic component that makes the drinks "tart." The carbonation and additional corrosiveness come from carbonic acid, which is formed from carbon dioxide and water.

Hydrochloric acid (HCl) is a common acid and is sometimes referred to as muriatic acid. It is used for several purposes, but it is commonly used to clean bricks, concrete, and adjust the pH balance in pools. Out of the acids discussed in this section, it has the highest vapor pressure. It is usually found in a 30–36 percent concentration range, but it varies with its use. Brick and concrete cleaner is usually 32 percent hydrochloric acid. With many acids, but most commonly associated with HCl, is the **Baumé scale**, which is an indication of the density of the acid. The density of the acid is also its specific gravity, which is measured with a hydrometer. With most acids water is a component, and the amount (density) of the acid in the water can vary. The Baumé scale provides an indication of the density in degrees Baumé. **TABLE 3-2** provides some information related to varying densities of HCl.

Nitric acid (HNO₃) is an acid that is occasionally seen in emergency response and has some additional hazards

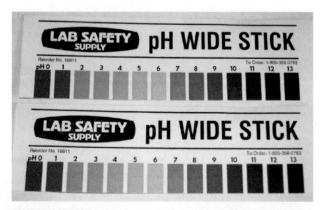

FIGURE 3-1 This is an example pH scale for pH testing paper that tests for pH levels 0–13.

Courtesy of Christopher Hawley.

TABLE 3-2 Properties of Hydrochloric Acid

Percent HCl	Degrees Baumé	pH	Vapor Pressure	Boiling Point
10	6.6	0.5	10.5 mmHg	226.4°F (108°C)
20	13	0.8	14.63 mmHg	217°F (103°C)
30	19	1	15.98 mmHg	194°F (90°C)
32	20	1.1	27.98 mmHg	183.2°F (84°C)
38	23	1.1	212.27 mmHg	118.4°F (48°C)

© Jones & Bartlett Learning

besides being corrosive. Nitric acid is also an oxidizer, which presents an additional hazard. The percentages of nitric acid vary, but the typical concentration is 60–70 percent. Like sulfuric acid, there is a fuming version of nitric, and in this case, there are two: White fuming nitric acid and red fuming nitric acid. There are many materials that produce a white vapor cloud, but there are very few chemicals that produce a red vapor cloud. Red fuming nitric acid has at least 10 percent dinitrogen tetroxide, which breaks down to nitrogen dioxide and is responsible for the red vapors. Red fuming nitric acid and bromine are the only two common chemicals that produce a red vapor cloud. **FIGURE 3-2** shows a red fuming nitric acid cloud coming from a reaction. White fuming nitric acid is more than 98 percent nitric acid and has very little nitrogen dioxide or dinitrogen tetroxide, less than 0.5 percent. In most cases, nitric acid is stored in a glass container as it is very corrosive to metals. If stored in a metal container, there may be a small amount of hydrofluoric acid (HF) in the nitric acid. The HF acts as an inhibitor and keeps the nitric from attacking the metal.

LISTEN UP!

An inhibitor is a material that keeps another material from initiating a reaction. It is commonly found in materials that can polymerize.

Hydrofluoric acid (HF) is another acid that is not commonly encountered but is seen on occasion. Like nitric it has some additional hazards. Hydrofluoric acid has a low pH but is considered a weak acid. It is, however, very reactive to glass and is commonly used to etch glass. One of the other hazards to humans involves the effect HF has on calcium, as it has a strong affinity for calcium. If someone gets HF on his or her skin, there

FIGURE 3-2 Red fuming nitric acid and bromine are the only two common chemicals that produce a red vapor cloud. When you see a cloud such as this, chemical protective clothing is required.
Courtesy of Maryland Department of the Environment Emergency Response Division.

may only be a slight burn or reddening of the skin, but below the skin, the HF is attacking the surrounding bones. In low concentrations, there may be a delayed effect on the skin, and someone contaminated with HF may appear fine but later suffer significant impact. Higher skin concentrations may produce excruciating pain and associated burns. The human cardiac system is reliant on a proper calcium balance, and exposure to HF can cause cardiac issues, including cardiac arrest. This is one example of the dose makes the poison, as small amounts of fluorine are beneficial to the human body. Water and some other products have fluoride added to benefit humans, but too much fluorine can cause fatal effects.

Hydrochloric, sulfuric, and nitric acids are also all called mineral acids, which is an early chemistry name that was used when minerals were distilled to make these acids. These acids are also inorganic as there aren't any carbon molecules in their makeup. A common organic acid is acetic acid (CH_3COOH), which has several different names, the most common being vinegar. Vinegar in the United States is typically a 3–7 percent concentration, but in Europe it is not uncommon to see 70 percent concentrations. It's for storage container size that the concentration is different. Another name is glacial acetic acid in which the "glacial" indicates purity. Glacial acetic acid is 100 percent pure acetic acid, and it has a freezing point of 62°F (17°C) and develops crystals near the freezing point.

Bases

At the opposite end of the pH scale are bases, which are also called alkali or caustics. They have equal hazards and risks as their acidic counterparts. The most common base is sodium hydroxide (NaOH), in which the formula exhibits the usual base identifier of having "OH" as part of its makeup. The OH is known as the hydroxyl component. Any compound with hydroxide as part of its name is probably a base. The most common base, sodium hydroxide, is also known by the common name of lye and is used as a drain cleaner. It is found in solid (crystal) and liquid forms and is readily available. It is used for cleaning products and is sometimes used in soap. Potassium hydroxide (KOH) is comparable to sodium hydroxide and used more in the industrial setting.

Ammonia (NH_3) is a base and is also part of the **amines** group, which is a group of compounds with nitrogen (N) as part of its makeup. Ammonia is very corrosive and in pure form is a gas known as anhydrous ammonia. When anhydrous ammonia is transported, it is a liquefied gas, which is liquefied by pressure as there isn't any water in the tank. If there is a breech in the tank, there will be a liquid release, which rapidly transforms to a vapor. In a nonpressurized liquid form, it has several names, including aqua ammonia and aqueous ammonia. When ammonia is in a solution of water (H_2O), it becomes ammonium hydroxide, the molecular formula of which is expressed as NH_4OH.

FIGURE 3-3 This shows a release of anhydrous ammonia.
Courtesy of David Binder.

Anhydrous ammonia is commonly used in industrial processes, refrigeration systems, and as a fertilizer (**FIGURE 3-3**). Ammonia is an interesting material and exhibits several hazards including corrosiveness, toxicity, reactivity, and flammability. The Department of Transportation (DOT) requires that when anhydrous ammonia is transported it is placarded and labeled as a nonflammable gas, which doesn't seem to fit with the listed hazards. The DOT has specific definitions for toxicity and flammability, although exhibiting these characteristics does not meet the specific DOT definitions. It is important to note that the DOT is concerned with regulating materials while in transportation, so some of the conditions and hazards that ammonia exhibits are outside their regulatory purview. Some other amines include methylamine, dimethylamine, and trimethylamine. Amines are known for their pungent irritating odors, which are detected at very low levels by humans. The amines odor is sometimes described as a fishy odor, as the odor you smell from rotting fish comes from two amines. The two amines produced by rotting flesh are cadaverine (1, 5-diaminopentane) and putrescine (1,4-diaminobutane), which contribute to the pungent odor. One study reports the **odor threshold** for ammonia is approximately 2.6 ppm, and the level of irritation is between 33 and 62 ppm.

When dealing with a corrosive response, one of the common methods of mitigating the release is to neutralize the corrosive. One thought is that water can be used to dilute, thereby neutralizing the spill. This presents two big issues: The mixing of water (chemical reaction) and the runoff from the reaction. Corrosives and water can be a dangerous combination. One should never add water to a corrosive as it presents great risk; there may be some spattering and heat generation.

If you had 1 gallon (3.8 liters) of an acid with a pH of 0, you would need 10 gallons (38 liters) of water to move the material to a pH of 1. To move it to a pH of 2, you would need 100 gallons (380 liters) of water, and to change the pH to 6, it would require 1 million gallons (3,785,412 liters) of water, all for a 1-gallon (3.8-liter) spill.

Chemically neutralizing a corrosive spill is the better choice, but even that can present some issues. When neutralizing a strong acid, you should use a weak base to perform the neutralization. A street method of calculation for neutralization states you would need more than 8800 pounds (3629 kg) of potash (potassium chloride or potassium carbonate) to bring 1000 gallons (3800 liters) of 50 percent sulfuric acid to neutral, which is more realistic for you to control than 1 million gallons (3,785,412 liters) of runoff. Soda ash (Sodium carbonate) is also commonly used to neutralize acid spills. To neutralize base spills, a weak acid is used, such as citric or acetic acids. Hazardous materials teams should research their jurisdiction for potential sources for these materials. Any industrial facility that does their own wastewater treatment will have some types of neutralization chemicals on hand. Although probably not a common thought, food and flavoring factories have large quantities of citric or acetic acid in addition to baking soda (sodium bicarbonate) for acid spills. Hazardous materials teams should have a list of potential resources such as neutralizing agents. These examples are for your information only; one should always consult a chemist about the neutralization of corrosives and can be very dangerous.

SAFETY TIP

If your skin and/or eyes are exposed to a corrosive material, they should be immediately flushed with large quantities of water. This flushing should continue for at least 20 minutes uninterrupted. Some corrosive materials can cause immediate blindness and skin burns, and water should be used to prevent further injury. It is important that at least 20 minutes of flowing water be continuously applied to the contaminated area. Hose lines, showers, or any source of water is critical to reducing the potential injury.

Hydrochloric acid (HCl) is commonly found at 20–25 percent and is known as muriatic acid (brick cleaner). It is available commercially at 32–37 percent. Removing relative concentration from the equation and

using only pH strength, we can look at the corrosives that have potential for immediate harm. Chemicals with a pH of less than 2 or more than 11 present a significant risk for injury to humans. Not all contact with material with a pH of 0 causes immediate skin damage. In many cases, the skin will only be irritated. In some cases, though, as with oleum, sodium hydroxide, or potassium hydroxide, the contact results in a burn within minutes and requires quick washing to prevent further damage. Having hydrochloric or sulfuric acid splashed on you results in skin irritation and burns after a few minutes, but quick washing can minimize or prevent the damage. Time is another factor in dealing with corrosives, as after a period of time a material with a pH below 2 or above 11 will cause damage to skin. The expression of time in this instance, however, is in hours. However, any corrosive in the eyes can cause immediate damage and possibly a permanent loss of vision. Examples of a common corrosive are Pepsi™ and Coke™, which have a pH of 2.5, which we can drink, but over time can cause damage to skin or painted surfaces, such as on a car.

SAFETY TIP

Any corrosives in the eyes can cause immediate damage and possibly a permanent loss of vision.

Methods of pH Detection

pH paper, often incorrectly referred to as litmus paper, comes in several different forms, but the most common is 1/4-inch rolls that have a pH range of 1–12 or 0–13. Also available are individual strips that cover the range of 0–14, as well as individual numbers. The multirange paper is the easiest and quickest to interpret. Individual strips with a multitude of colors to match are more difficult to correctly interpret. It is advisable to test your pH paper for its effectiveness, as it varies from brand to brand among the color ranges. For example, the pHydrion™ 0–13 range is much quicker to respond to gases than the 1–12 range paper.

The most common paper has a range of 0–13 and has one-step increments based on a color change. Paper is available in smaller ranges (1–3, for example) and with increments such as 1.1, 1.2, 1.3, and so on. Multicolor strips are made that offer the same ranges as the paper rolls and offer some additional reliability. It is not generally important to have pH paper that tells the pH is 1.3

when a less expensive roll lets us know that it has a pH of 1. For emergency response, it is not necessary to have the exact pH of a spilled material, but responders do need to know if it is corrosive or neutral. It is also necessary to know whether the material has a corrosive vapor or is just a liquid corrosive.

One method to test the pH of solid materials such as concrete, wood, walls, and even skin is a pH mechanical pencil. One type of pencil manufactured by Hydrion has a pH range of 0–13 and indicates the pH of a material via a color change. To use the pencil, you write a stripe on the item to be tested and then wait 15 seconds for the color change.

Another method of determining pH of materials is to use a pH probe, which is commonly used in labs. These probes start at $100 and run into several thousand dollars. They are not commonly used in emergency response, as the sensors need to be replaced regularly (6 months to a year), and they must be kept wet as they are ruined when dried out. They are kept wet in a buffer solution, which keeps the pH constant. A buffered solution is a liquid that has dissolved solids that resists changes in pH. They also require calibration using 2–3 buffer solutions.

The sensors must be calibrated before each use, so they must be maintained with the appropriate calibration solutions. The sensors are shock sensitive and are easily broken during the rigors of field use. The only time this type of device might be useful is when sampling many materials in a fairly controlled environment. However, the pH probe requires cleaning after each sampling prior to measuring another sample.

Because pH paper turns colors in the presence of corrosive vapors, it can be used to indicate corrosive atmospheres as well as liquids. Dry pH paper indicates the presence of corrosive vapor. Some teams will wet the pH paper prior to use with deionized water to check for liquids and vapors. Some pure acids do not indicate on dry pH paper, but this would be an unusual case. The multirange paper is more than sensitive enough for detecting vapor in the air when dry. For ammonia vapors, dry multirange pH paper indicates at 1 ppm. Using a setup as shown in **FIGURE 3-4** provides protection against corrosive gases and does not require handling. The X or T format allows for three or four liquid samples to be taken with one pair of forceps, but if vapors are present, all the paper may change color.

In some cases, the liquids you are sampling may present misleading readings on pH paper and may be misinterpreted. When reading pH paper, you not only need to identify the color change but also to interpret the leading edge of the paper. This is usually only a

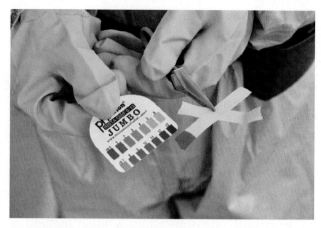

FIGURE 3-4 By using a T or an X formation, you can accomplish three or four tests with one pair of forceps.
© Jones & Bartlett Learning. Photographed by Glenn E. Ellman.

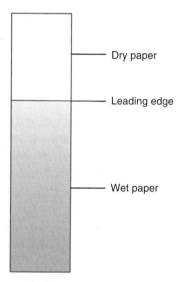

FIGURE 3-5 When the pH is reported to be between 4 and 8 and has a straight leading edge, this liquid is most likely a neutral substance.
© Jones & Bartlett Learning

concern when sampling liquids you think have a pH of 4–8. All other pH levels are extreme enough that their results usually are not misinterpreted. The problem is usually with hydrocarbons, which are neutral substances, but when sampling for pH, they may cause an erroneous reading of the pH paper. When multirange pH paper is wetted with a substance such as a hydrocarbon, it gives the impression that it has a pH of 4–8. The paper is just wet and really did not change. Two methods can be used to determine if the pH is accurate. If the leading edge of a material is indicating a pH of 4–8 and is straight, the material is most likely neutral (**FIGURE 3-5**). This test is not 100 percent but offers fair reliability. The other test, which is 100 percent accurate, only works with high vapor pressure materials. When sampling, stick the pH paper into the vapor space of the sample jar, and check for a change. If there is a change in the vapor space, then check the liquid's pH and record those results, as this material is a corrosive. If the pH paper did not change, but in the liquid you are interpreting the readings have a pH of 4–8, then record those readings and complete an evaporation check. With a pipette, take a small sample from the sample jar. Place one drop on a watch glass, start a clock, and watch the sample. If the sample evaporates in 5 minutes or less, then the material is neutral. A material that evaporates in 5 minutes has a high vapor pressure and would have changed the pH paper in the vapor space. If it did not, the pH reading was from wet pH paper. A leading edge on pH paper that is jagged, multicolored, and seems to be wicking through the pH paper is a corrosive and should be recorded as such (**FIGURE 3-6**). The pH of common materials is provided in **TABLE 3-3**.

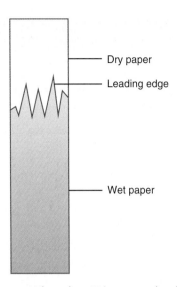

FIGURE 3-6 When the pH is reported to be between 4 and 8 and the leading edge is jagged, this substance is a likely corrosive.
© Jones & Bartlett Learning

TABLE 3-3 pH of Common Materials			
Material	**pH**	**Material**	**pH**
Orange juice	3	Sulfuric acid	0
Ammonia	13	Pepsi™	2–3
Vinegar (acetic acid)	3	Hydrochloric acid	0
Sodium hydroxide (lye)	13	Oleum	1
Acetone*	7	Gasoline*	7
Hydrofluoric acid	0–1	Diesel fuel*	7

*Materials such as hydrocarbons are technically not corrosive because they cannot produce/liberate hydrogen ions or hydroxide ions, but they indicate on pH paper. When pH paper is dipped into one of these materials, the wet paper appears to provide a pH reading, but experienced responders recognize the straight leading edge and report that the material is a hydrocarbon along with an accurate pH.
© Jones & Bartlett Learning

After-Action REVIEW

IN SUMMARY

Testing for corrosives is a very important activity, not only for responder safety but also for the preservation of the instruments. Corrosives are the second most released substance behind flammables and combustibles, so you must be comfortable with pH sampling. Practice with known materials, and you will become familiar with how common corrosives react.

KEY TERMS

Acid Material that has a pH of less than 7.

Alkali Material with a pH of greater than 7. Also known as a base or a caustic.

Amines Group of compounds with nitrogen (N) as part of its makeup.

Base Material with a pH greater than 7. Also known as an alkali or a caustic material.

Baumé scale An indication of the density of the acid.

Concentration The amount of a material in a certain volume.

Corrosive Material that can cause damage to skin or metal and is either acidic or basic.

Hydrolysis material The breakdown product(s) of a material in reaction with water.

pH An abbreviation for potential of hydrogen, power of hydrogen, percentage of hydrogen ions, and the negative logarithm of hydrogen ion activity.

pH paper Testing paper used to indicate the corrosiveness of a liquid.

Review Questions

1. What is the best method for testing for corrosive vapors in the street?

2. What does pH stand for?

3. Material that has a pH of 4 is a(n) _____.

4. Material with a pH of 11 is a(n) _____.

5. Material with a pH of 7 is _____.

6. The pH scale is known as what type of scale?

7. Material with a pH of 1 is _____ times more corrosive than material with a pH of 2.

Oxygen Level Determination

LEARNING OBJECTIVES

Upon completion of this chapter, you should be able to:

- Describe the importance of knowing the oxygen concentration
- Describe the methods used to identify the oxygen concentration
- Describe the common types of oxygen sensors

Hazardous Materials Alarm

An engine company and an ambulance are dispatched for an unconscious person in a fast food location. Upon arrival, there is an unconscious employee in the storeroom and two patrons who are not feeling well. They had been in an adjacent bathroom that is near the storeroom. The engine officer decides to evacuate the restaurant and quickly remove the victim to the outside. He calls for a hazardous materials assignment to assist with the investigation. After a few minutes of being outside and being on oxygen, the unconscious victim starts to improve. The hazardous materials team arrives, and the initial monitoring determines that there are high levels of carbon dioxide (CO_2). They find a CO_2 tank had leaked, filling the storeroom with CO_2 and flowing into the bathroom.

1. Where in your sampling protocol would you look for CO_2?

2. Is CO_2 a common material in fast food locations?

3. What if there had been a basement?

Introduction

After pH, oxygen is the next most important thing to sample for, as humans need it to survive, and some monitoring instruments need it to function correctly (**FIGURE 4-1**). Normal air contains 20.9 percent oxygen; below 19.5 percent, or **oxygen deficient**, is a health risk, and above 23.5 percent, or **oxygen enriched**, is considered a fire risk. If an oxygen drop is noted on the monitor, one or possibly more than one contaminant is present, causing the reduced oxygen levels, and another hazard (i.e., toxic, flammable, corrosive, or **inert**) is causing the oxygen-deficient atmosphere. In oxygen-deficient atmospheres, any combustible gas readings are also deficient and are not reliable. In an oxygen-enriched atmosphere, the combustible gas readings increase and are not accurate.

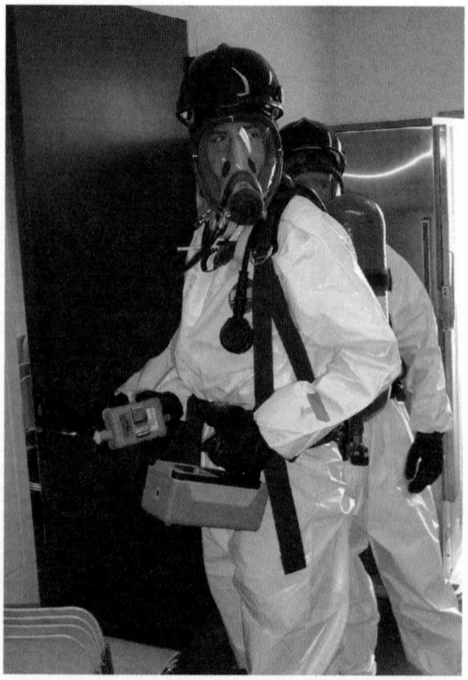

> **LISTEN UP!**
>
> Every 0.1 percent drop in oxygen on the meter was worth 5000 ppm of something else, which is a considerable amount of a potential toxic material in the air.

Oxygen-Related Incidents

Within Department of Transportation (DOT) regulations and in the world of chemistry, there is a group of materials known as oxidizers. The term "oxidizer" indicates a material's ability to give up oxygen molecules, or, in other words, release or produce oxygen. By itself, oxygen, which is expressed as diatomic oxygen or O_2, is not flammable. When something is burning and there is additional oxygen present, the burning rate is increased. Materials burn faster and at a higher temperature. With oxidizers, there is the potential for a fire or an explosion. Fire itself is defined as the rapid oxidation of a material

FIGURE 4-1 Monitoring for oxygen is important for human survival and to ensure the instruments are function normally.
Courtesy of Christopher Hawley.

during combustion with accompanying heat, light, and byproducts. To have a fire, there must be three basic elements: Fuel, oxygen, and a heat source. In reality, there is a fourth component, which is a chemical reaction. The group of chemicals known as oxidizers adds the oxygen component, and some will add the chemical reaction as well. Once a fire ignites it can become self-sustaining, and the reaction will continue until the fuel is consumed. One of the clues that a material may be an oxidizer is a portion of its chemical name. Chemicals that have peroxide, chlorate, perchlorate, nitrate, and permanganate as part of their name are probably oxidizers. The NFPA has four categories of oxidizers, which are outlined in **TABLE 4-1**.

Most oxidizers primarily aid in the burning rate of a substance, but there are some that present some unusual hazards. Those listed in class 3 and 4 are materials that are extremely hazardous. They react with and can ignite organic materials such as paper, wood, hydrocarbons (flammable and combustible), and powdered metals. They can also react with other substances that are oxidizers such as hydrazine, hydrogen, sulfur, and ammonia. If contaminants such as a powdered metal get into a container of an oxidizer, there is potential for a reaction, which can be a fire or an explosion.

Included as part of the oxidizer group but is a more dangerous subset is a group known as organic peroxides. These are compounds that have a double oxygen molecule, which is represented by the O-O symbol. It is the group of peroxides that are especially dangerous, depending on their concentration. By examining Table 4-1, you see that hydrogen peroxide increases in danger as the percentage of hydrogen peroxide increases. Normal household hydrogen peroxide is typically 3–5 percent with the majority being water, and at that percentage, it does not present any oxidizing

TABLE 4-1 Common Oxidizers and Classification Categories

Class 1: An oxidizer than does not moderately increase the burning rate of combustion.

Aluminum nitrate	Ammonium persulfate	Calcium chlorate
Hydrogen peroxide (8–27.5%)	Nitric acid <40%	Perchloric acid (<50%)
Potassium dichromate	Potassium nitrate	Sodium nitrate
Sodium nitrite	Sodium perchlorate	Sodium persulfate

Class 2: Moderately increase the rate of combustion.

Calcium chlorate	Calcium hypochlorite (<50%)	Chromic acid
Hydrogen peroxide (27.5–52%)	Magnesium perchlorate	Nitric acid (>40% but less than 86%)
Potassium permanganate	Sodium permanganate	Sodium peroxide

Class 3: Severely increase the rate of combustion and may cause sustained and vigorous decomposition when exposed to heat or encounters a combustible material.

Ammonium dichromate	Hydrogen peroxide (52–91%)	Fuming nitric acid (>86%)
Perchloric acid (60–72%)	Potassium bromate	Potassium chlorate
Sodium chlorate	Sodium chlorite (>40%)	Sodium dichloroisocyanurate

Class 4: Can ignite or explode in contact with some materials or through heat, shock, or friction.

Ammonium perchlorate (>15 microns)	Ammonium permanganate	Hydrogen peroxide (>91%)
Perchlorate acid (<72.5%)	Tetranitromethane	

risk. At mixtures of 91 percent or more, there is a great risk with this material. Some organic peroxides are susceptible to violent reactions to heat, shock, and cross contamination or the mixing with other materials. The list of potential materials that can react is extensive, and hazardous materials teams should consult with a chemist for assistance.

Monitoring for Oxygen

In oxygen extremes (deficient/enriched), there is a major problem because a considerable amount of some gas (may include oxygen) is present. When there is a drop in oxygen, the amount of occupying gas can be calculated with some known information.

Note: For the sake of this scenario, the calculations are rounded off. In air we have 79 percent nitrogen (N^2), which equates to 790,000 ppm, and roughly 21 percent oxygen (O_2), which equates to 210,000 ppm oxygen, as shown in **FIGURE 4-2**. Using a rough rule of fifths, the makeup of air is four-fifths N_2 and one-fifth O_2. If it is assumed there is a contaminant that will drop the O_2 content by 5000 ppm, it can then be assumed that the N_2 will drop by 20,000 ppm as the entire content of the air will be affected (four times as much contaminant in air to drop N_2), so the total amount of contaminant in the air is 25,000 ppm or 2.5 percent by volume. In this example, the oxygen level would read 20.4 percent (using 20.9 percent as our starting point), which also means that every 0.1 percent drop of oxygen on the meter was worth 5000 ppm of something else—a considerable amount of a potential toxic material in the air. A drop in oxygen that goes from 20.9 percent to 20.7 percent can be significant, and responders should wear self-contained breathing apparatus (SCBA) as a minimum. A drop on the oxygen meter of 1 percent, from 20.9 to 19.9 percent, would indicate that 50,000 ppm of something has entered the atmosphere, well above immediately dangerous to life or health (IDLH) toxicity levels for many chemicals. One warning, though, is that when new oxygen sensors are turned on, the readings will fluctuate and, in some cases, provide erroneous readings until the electronics are warmed up. The meter may indicate its readiness, but until the unit is warmed up and running for 7–10 minutes, the reading will fluctuate, particularly in cold winter conditions. Another problem with oxygen sensors is that some manufactures try and correct the fluctuation of the sensors by restricting the changing readings, known as banding. They force the sensor to read 20.9 when the actual reading could vary from 20.8 to 21.0. Oxygen sensors have not yet developed enough to make them stable for accurate readings, which is a challenge. When an oxygen sensor drops or increases, it is a significant change, which can present a risk to responders.

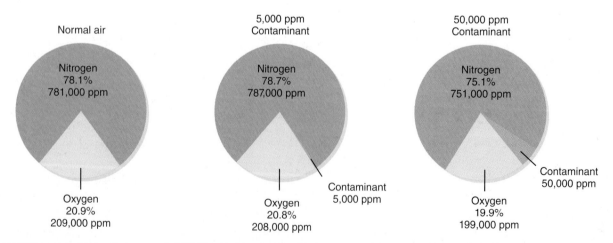

FIGURE 4-2 Normal air contains 20.9 percent oxygen, 78.1 percent nitrogen, and 1 percent a variety of other gases. We use only oxygen and nitrogen for this example, because the 1 percent of the other gases further complicates the explanation. Normal air is represented by **A**; when the atmosphere is contaminated, both the oxygen and nitrogen are reduced, as shown in **B** and **C**. In **B** the contaminant drops the oxygen to 20.8 percent, and in **B** the level of oxygen is 19.9 percent. In both cases the oxygen sensor would not alarm, but a contaminant at 5,000 ppm or 50,000 ppm has great potential to cause serious injury or death.

FIGURE 4-3 An example CO_2 tank found in fast food restaurants.
Courtesy of Christopher Hawley.

CARBON DIOXIDE INCIDENTS

As is discussed in Chapter 14 on sick buildings, CO_2 is an important gas to sample for early in hazardous materials responses. Chapter 14 also describes the typical installation of a CO_2 tank. There is another reason to place CO_2 higher on your sampling priorities, and that is due to the proliferation of CO_2 in commercial buildings. Most fast food locations, restaurants, bars, convenience stores, and other buildings use large quantities of CO_2, usually found in liquidized gas containers (**FIGURE 4-3**). Any business that dispenses soda will most likely have a CO_2 cylinder or a CO_2 dewar.

The properties of CO_2 are:

- Exists as a solid, liquid, and gas
    - Typically, a gas is stored as a refrigerated liquid.
- Slightly acidic ~3
- Vapor density 1.5
- Melting point −70°F (−57°C)
- Boiling point −109°F (−78.5°C)
- Ionization Potential (I.P.) −13.77
- IDLH 40,000 ppm, 4 percent by volume
- Time weighted average (TWA) 5000 ppm, 0.5 percent by volume
- Short-term exposure limit (STEL) 30,000 ppm, 3 percent by volume

In recent times, there have been fatalities and injuries due to incidents at fast food locations due to CO_2

incidents. They were becoming significant enough that Occupational Safety and Health Administration (OSHA) issued a safety alert about CO_2 installations. Here are some of the recent incidents:

- Phoenix, Arizona, fire and EMS crews were dispatched to assist with a female who had fallen and was feeling ill. Upon arrival at a fast food location, crews found a pregnant female who, upon reaching the top of the basement steps, fell and was not feeling well. While other crewmembers were treating the patient, two other crewmembers started to explore the basement. On the way down, they started getting light-headed and felt ill. They were able to self-rescue and make it back up the steps to report their findings. The officer evacuated the building and requested a hazardous materials assignment. When the hazardous materials crews entered the building, they found decreased oxygen levels and also received a methane reading. They found that a recently filled CO_2 tank in the basement had leaked, creating an oxygen-deficient atmosphere. The CO_2 alarm that had been installed to warn employees of potentially toxic situations was shut off. The low oxygen response on the instruments was expected, but the lower explosive limit (LEL) reading would not normally be found. The responders were using a multigas device with a thermal conductivity (TC) LEL sensor. This sensor reacts to thermal changes, of which CO_2 will cause a reaction. See Chapter 5 for more information on this sensor.
- Emergency response crews responded to a fast food location for two unconscious persons. Upon arrival they found two women in the restroom, one unconscious and one in cardiac arrest. They were removed and medically treated. The initial assumption was that cleaning chemicals had caused a toxic environment affecting the women. When the news reports made the national news, crews from the Phoenix, Arizona, Fire Department called to let them know about their CO_2 incident. Crews went back to the restaurant and found the CO_2 fill line, which ran through the bathroom, was leaking and was the cause.
- A delivery driver attempted to deliver CO_2 to a fast food location and found the door where the tank was located locked. One of the employees offered to jump over the wall that encased the tank and hook up the hose. The driver was not able to get a response from the employee and also jumped over the wall. Another employee later noticed that the original employee and the truck driver were missing. They located the key and opened the door where they found them both

dead. The CO_2 fill line was leaking and had filled the confined space with CO_2.

- Hazardous materials crews were called to assist with emergency medical services (EMS) crews that were treating some sick persons. The occupancy was a typical office, and no apparent cause was evident. Hazardous materials crews located a desktop soda-dispensing machine that had a CO_2 cylinder next to it. Crews checked the CO_2 levels, which were elevated. Crews asked the patients about the CO_2 cylinder. The occupants reported that they had changed the cylinder, as it was empty. Crews checked the recently installed cylinder and found it empty as well. The system was leaking CO_2, which made the occupants sick due to their long-term exposure.

Given the increase in the number of these CO_2 containers in use today, emergency responders should always anticipate a release when responding to a location that has one of these containers.

Oxygen Sensors

There are three types of oxygen sensors: Two versions of an electrochemical lead wool sensing technology and a **solid polymer electrolyte (SPE)** oxygen sensor (**FIGURE 4-4**). The SPE sensors are attractive because they have a longer life than the lead wool sensors and can last up to 5 years (**FIGURE 4-5**) and (**TABLE 4-2**). The SPE sensor, also known as an oxygen pump sensor, draws in the oxygen through a capillary. The oxygen travels into the sensor housing and is reduced to water at the sensing electrode. In the bottom half of the sensor, water is oxidized into oxygen, and protons are released. The protons travel across the SPE membrane,

FIGURE 4-4 Typical oxygen sensors.
Courtesy of Christopher Hawley.

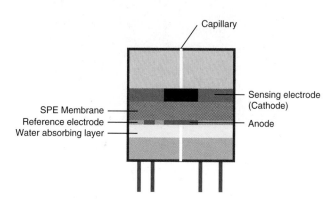

FIGURE 4-5 Makeup of a typical SPE sensor.
© Jones & Bartlett Learning

and the oxygen is vented through a second opening on the bottom side of the sensor. The advantage of the SPE sensor is that its operation does not require a limited supply of materials. The one drawback is its power consumption, but RAE Systems by Honeywell (an SPE sensor manufacturer) has developed a work-around to combat this issue. Oxygen sensors are linear up to

Name/Type	Expected Life	T90 Response	T97 Response	Range
4OxLL/Electrochemical	5 years	<15 secs	<35 secs	0–25%
7OX-V/Electrochemical	2 years	N/A	<15 secs (T95)	0–25%
RKI Eagle O_2		<30 secs		0–40%
Dräger XS EC O_2	3 years	<25 secs		0–25%
Dräger XS 2 O_2	2 years	<20 secs		0–25%
Dräger XS O_2 microPac	5 years	<32 secs		

TABLE 4-2 Oxygen Sensor Specifications

30 percent oxygen content. Partial pressure sensors are linear beyond that and up to 100 percent.

SAFETY TIP

When oxygen sensors are turned on, the readings can fluctuate and in some cases provide erroneous readings until the electronics are warmed up. Typically this takes 2–3 minutes.

Of the **lead wool sensors**, the more common is a **capillary pore sensor**, with the other being a diffusion **membrane sensor**, also known as a partial pressure sensor. Both act like a battery and are sometimes referred to as a fuel cell sensor. Internally they have at least two electrodes in a basic solution, typically potassium hydroxide (**FIGURE 4-6**). One of the electrodes will commonly be lead. The basic solution, known as an **electrolyte**, is in a slurry or gel-like form. Inside the cell is a small amount of lead wool, which is the center of the chemical reaction. When oxygen enters the sensor housing, an oxidation reaction generates electrical activity, generating an electrical charge. The electrical activity is measured and converted to a percentage-of-oxygen reading. The lead wool is the weak link in this process. During the oxygen reaction, the lead wool converts to lead oxide, and when all the lead is converted to lead oxide, the sensor ceases to function. This sensor is always functioning whether the device is on or not. Once the sensor is removed from an inert gas tight container, it is reacting, and the lead oxide is being produced reducing the life of the sensor.

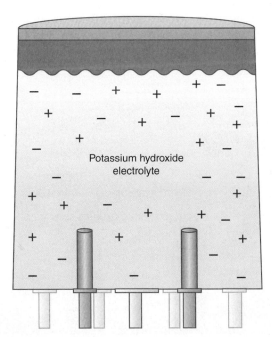

Potassium hydroxide electrolyte

FIGURE 4-6 Makeup of a typical lead wool sensor.
© Jones & Bartlett Learning

The capillary sensor is similar but has a small tube (a capillary pore) through which the oxygen travels into the cell. These types of sensors last only 2 years because the lead wool is continually being converted to lead oxide. The capillary sensor is not subject to changes in atmospheric pressure like the diffusion membrane sensor. If you calibrate a diffusion membrane at sea level and then transport it to a higher elevation, in the thousands of feet, the readings can differ. In outside air at sea level, the atmospheric pressure is 14.7 psi, and at 5000 feet, the oxygen content is 20.9 percent, but the atmospheric pressure will be 12.45 psi, which is reducing the amount of oxygen reaching the diffusion membrane. With the lead oxide sensors, the instrument can be brand new and never turned on, but the oxygen sensor degrades every day. In multiple gas detection devices that house a variety of sensors, the oxygen sensor is the one most frequently replaced. The amount of oxygen that hits the sensor determines its ultimate life span. In addition to strong oxidizing substances, primarily over IDLH levels can also degrade the sensor and impact the readings. With chronic exposure, carbon dioxide, which is corrosive, can degrade the sensor, so chronic or long-term exposure to carbon-dioxide-laden environments should be avoided. Chlorine and ozone are strong oxidizing substances and also shorten the sensor's life.

Other items that affect primarily the diffusion oxygen sensors are heat, cold, humidity, and atmospheric pressure. Because the electrolyte is a semifluid, cold weather causes the sensor to react slowly and, at some point, quit functioning. Hot weather has the opposite effect, and it can also damage the sensor. The sensor can compensate for normal fluctuations in temperatures but has trouble with extremes. The optimal temperature range for operation is between 32°F and 120°F (0 and 49°C). At temperatures between 0°F and 32°F (-18°C and 0°C), the sensor slows down. The conversion of lead-to-lead oxide causes expansion in the sensor housing because lead oxide takes up more space than lead. The sensor is designed to accommodate this increase, but when the temperatures get below 0°F (-18°C), the electrolyte can freeze, causing an expansion in the cell and permanently damaging the sensor by freezing. If the sensor is old, it may rupture due to the expansion of the lead oxide. On rare occasions, moisture can block the capillary tube, causing the SPE sensor to malfunction. Humidity's impact varies with the temperature and the humidity level. With one electrochemical sensor at 32°F (0°C), there is little change to the sensor reading even at 100 percent humidity. At 68°F (20°C), the sensor drops to 20.8 percent at 30 percent humidity. At 60 percent humidity, the sensor reads 20.6 percent oxygen. Near 100 percent humidity, the sensor will drop to approximately 20.5 percent. At 86°F (30°C), the drop is even more dramatic. The sensor will read 20.6 percent at 20 percent humidity and will read 20 percent at 60 percent humidity. At 100 percent humidity, the reading would be 19.4 percent. Gradual changes in atmospheric pressure usually do not impact the oxygen sensors, but a rapid change may produce a change. Some sensors, if shipped on an airplane to a different elevation, may require a 1-hour acclimation period at a new elevation. When you check a manufacturer's information on lead oxide sensors, it states that the sensor may take 20 minutes to fully warm up. Luckily, the oxygen sensor can be fresh air set, and some mixed gas calibration gas does use oxygen as one of the gases. It is best to calibrate the instrument in the conditions that will most likely be found during the air monitoring operations.

LISTEN UP!

Exposures that hurt oxygen sensors are oxidizers above their IDLH and chronic exposures to carbon dioxide (CO_2) and strong oxidizing materials, such as chlorine and ozone.

After-Action REVIEW

IN SUMMARY

Measuring oxygen is of prime importance, not only because we need oxygen to survive, but because most monitors also require oxygen to function. The fact that an atmosphere may be oxygen deficient provides significant clues as to the potential for large amounts of a toxic or asphyxiating substance being present. Being in an oxygen-enriched atmosphere presents a greater fire risk and usually means that free oxygen or an oxidizer is involved in a chemical reaction.

KEY TERMS

Capillary pore sensor A sensor that uses a small tube to carry the gas into the sensor housing.

Electrolyte Material used to conduct electrical charges inside a gas sensor.

Inert A chemical that is not toxic but displaces oxygen.

Lead wool sensors A sensor that uses lead wool in the detection of oxygen.

Membrane sensor A sensor that relies on the gas moving through a membrane on top of the sensor.

Oxygen deficient Oxygen level below 19.5 percent.

Oxygen enriched Oxygen level above 23.5 percent.

Solid polymer electrolyte (SPE) A sensor that uses a membrane to diffuse oxygen through water to measure the oxygen content. Also known as an oxygen sensor.

Review Questions

1. What is the normal percentage of oxygen in air?

2. What amount of oxygen in air makes it oxygen deficient?

3. What amount of oxygen in air makes it oxygen enriched?

4. What is the risk of an oxygen-enriched atmosphere?

5. Lead oxygen sensors operate much like what type of device?

6. When do electrochemical oxygen sensors begin to detect oxygen?

7. What does a 0.1 percent drop in oxygen mean?

CHAPTER 5

Flammable Gas Detection

OUTLINE

LEARNING OBJECTIVES

Upon completion of this chapter, you should be able to:

- Identify the methodology behind flammable gas sensors
- Describe the impact of lower explosive limit (LEL) levels for a variety of gases and vapors
- Describe the advantages and disadvantages of basic flammable gas sensors

Hazardous Materials Alarm

A hazardous materials assignment was dispatched to an industrial facility. In the previous year, the hazardous materials team had responded to the facility for an underground nitrocellulose tank that was leaking. Nitrocellulose is also known as gun cotton, the main component of Sterno™. That event was not particularly significant, as it involved the nitrocellulose bubbling up out of the ground into an adjacent parking lot. The next call also involving nitrocellulose would be much more significant. Since the underground tank leak, the facility was getting nitrocellulose in 300 gallon totes, which were stored in an adjacent warehouse that also housed all their hazardous waste. These drums of flammable waste were stacked from floor to ceiling, in addition to many other totes in the building. This building was adjacent to the main manufacturing facility and was just under the highway overpass. When the hazardous materials crews arrived, they found that 300 gallons of nitrocellulose had been spilled out onto the warehouse floor, coating pallets and other drums and totes. Upon arrival, the corrected LEL readings were 90 percent of the LEL—a cause for concern. Because this liquid was flammable, foam was applied, which brought the LEL down to 85 percent. The hazardous materials team was concerned, as it knew from the previous experience that nitrocellulose, usually stabilized with ether, alcohol, acetone, and water, becomes a shock-sensitive explosive as it dries out. The team also knew that if it did ignite, it would be nearly impossible to extinguish without a large quantity of purple K extinguishing agent. There were some discussions about the cleanup and the contractor that would be needed to complete the job. The initial contractor balked at entering the atmosphere at 85 percent of the LEL, but the longer crews waited, the worse the explosive hazard would be, so cleanup was quickly initiated along with stabilization. It is always recommended to work in levels below 25 percent or 10 percent of the LEL, but this is not always possible.

1. What is the most common gas that LEL sensors are set to read?

2. What does a corrected reading mean?

3. What is a dangerous level for an LEL?

Introduction

Flammable gas indicators, which are also known as **combustible gas indicators** (CGIs), have been used by the fire service and industry for many years. The term "combustible gas indicator" does not accurately describe what they detect. In reality, they only measure flammable gases not combustible gases. A more accurate name for these meters is flammable gas indicators or an LEL-indicating meter. The actual sensor that does the detecting should be called an **LEL sensor**, the term that is used throughout the remainder of this text.

Most of the new **LEL meters** are used to measure the lower explosive limit (LEL) of the gas for which they are calibrated (**FIGURE 5-1**). Instruments vary in

FIGURE 5-1 A typical LEL monitor that is part of a four-gas detection device.
Courtesy of Christopher Hawley.

size, with some that can fit in the palm of your hand (**FIGURE 5-2**). LEL meters are used to identify the fire risk in the risk-based response system. Most LEL meters are calibrated for methane (natural gas) using methane or pentane gas. When calibrated for methane, the LEL sensor will read up to the LEL; some new units shut off the sensor when the atmosphere exceeds the LEL to protect the LEL sensor. This is an important consideration because the longer the sensor is exposed to an atmosphere above the LEL, the faster it will deteriorate. LEL sensors are available that read above the LEL by reading in percent by volume. These sensors are known as volume or concentration devices. Landfills and some industrial applications need LEL meters that read 100 percent by volume because many times these applications have methane in quantities that greatly exceed the LEL, and they need to monitor the high levels. Although having this capability has advantages, it really does not matter whether the atmosphere is 100 percent of the LEL or 70 percent by volume. Once the LEL is exceeded, there is a fire hazard. Responders' tactics will not change because the volume LEL meter reads 74.5 percent by volume and the LEL meter reads 100 percent of the LEL.

LISTEN UP!

The term "combustible gas indicators" does not accurately describe what they detect. Responders should use the term "flammable gas indicator" or "LEL sensor."

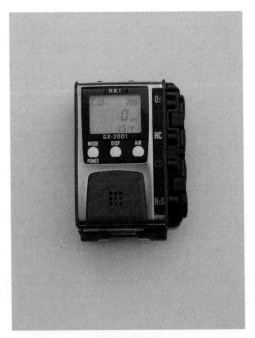

FIGURE 5-2 A miniature four-gas instrument that detects flammable gases, CO, H₂S, and oxygen content.
Courtesy of Christopher Hawley.

The LEL meter reads up to 100 percent of the LEL, which for methane is 5 percent volume in air of methane. Thus, if an LEL meter calibrated for methane reads 100 percent, the actual concentration by volume in air of methane is 5 percent. If the LEL meter reads 50 percent, then the concentration for methane is 2.5 percent by volume in air.

SAFETY TIP

LEL meters are used to identify the fire risk in the risk-based response system.

Any flammable gas sample that passes over the sensor causes a reaction and how much of a reaction depends on the gas. Each LEL meter comes with a correction factor for other gases. The exact number of other gases referenced varies from manufacturer to manufacturer. Correction factors are discussed in Chapter 1.

SAFETY TIP

If you obtain a reading of 1 percent on your properly calibrated LEL meter at the door to a building, you may have a problem; there is a flammable gas/vapor present.

Both OSHA and the Environmental Protection Agency (EPA) have established action levels, shown in **TABLE 5-1**, that provide a safe level expressly because of the relative response curve problem. The OSHA values are from the confined space regulation (29 CFR 1910.146) and are only applicable in confined spaces, although many response teams have adopted those values for standard hazardous materials work. They have factored the various response curves in and provide guidelines that can be followed.

The unfortunate thing about the EPA action guidelines is that they were originally designed for hazardous waste sites (which are usually outside), and they do not apply to a lot of what the fire service does, as in one case described in the nitrocellulose case study. Some factors need to be established that would impact these action guidelines, because in some cases, a 26 percent reading may require action.

If you respond to a building for a natural gas leak and the meter reads 26 percent, this situation should be considered dangerous. The determining factor is where this reading was obtained. If it was obtained at the front door and you have not identified the source, then you

TABLE 5-1 OSHA and EPA Monitoring Action Levels

Atmosphere	Level	Agency	Action
Combustible gas	<10% LEL	EPA	Continue to monitor with caution.
	10–25% LEL	EPA	Continue to monitor, but use extreme caution, especially as higher levels are found.
	>10%	OSHA	Requires evacuation of the confined space
	>25%	EPA	Explosion hazard: Evacuate the area.
Oxygen	<19.5%	EPA	Monitor with SCBA; CGI values are not valid.
	19.5–25%	EPA	Continue monitoring with caution; SCBA not needed based on O_2 content only.
	19.5–23.5%	OSHA	Continue to monitor.
	>25%	EPA	Explosion hazard: Withdraw immediately.
	>23.5	OSHA	Explosion hazard: Withdraw immediately.

© Jones & Bartlett Learning

may be in serious trouble. If the highest reading you obtain is 26 percent and you have located the source of the leak (and hopefully shut the gas off and have ventilated the building), you are in a relatively safe situation. To have a fire, the meter needs to read 100 percent, so in each instance you were 74 percent in the good range. If there is no other life hazard, at 26 percent you have no reason to be in the building and should evacuate. When dealing with unidentified situations, which include natural gas leaks as they are unidentified until you confirm the source, you should consider a reading of 1 percent significant. If you obtain a reading of 1 percent on your properly calibrated LEL meter at the door to a building, you may have a problem, and there is potential that somewhere in that building the gas will be near the LEL. Keep in mind that 1 percent of the LEL equates to about 100–200 ppm, which is a significant amount of a potentially toxic material. If the gas was methane, then the amount would be 500 ppm. When dealing with life hazards, we really should not quantify safe or unsafe because our obligation is to protect the public. Chapter 10 discusses risk and making life hazard and rescue decisions, but the reality is that there cannot be a fire until the LEL meter reads 100 percent with a corrected reading. Even when it reads 100 percent, people should be rescued as rapidly as possible. Anytime you

move people they are in danger; when roads are shut down, people are in danger for additional accidents.

People panic and do not follow directions well. People will do everything in their power to escape a real or perceived threat. Sometimes the best course of action is to remain sheltered in place. When there is no public life risk, at 10 percent LEL, first responders should evacuate the building. When dealing with unidentified materials, some additional considerations come into play, specifically correction factors. The worst response curve for LEL sensors is a 5, which is rarely encountered, as all the others are generally below 2.6, which we will round up to 3 for ease of computation and to add a layer of safety. With a factor of 3 and an LEL meter reading of 33 percent, the actual atmosphere could be 99 percent, so when dealing with an unidentified gas, the action level should be 33 percent. By using other detection devices and identifying the chemical family, you may be able to increase this number to the lowest range for that family.

SAFETY TIP

When there is no public life risk, at 10 percent LEL, first responders should evacuate the building.

Flammable Gas (LEL) Sensor Types

The basic principle of most LEL sensors is that a stream of sampled air passes through the sensor housing, causing a heat increase, increasing the resistance in an electrical circuit, which when balanced by a known fixed resistance causes a reading on the instrument. The infrared sensor, thermal conductivity, and laser sensors are the only ones that do not follow that principle. There are six types of LEL gas sensors, and when purchasing or using a monitor, it is important that responders know the sensor type. The types are **Wheatstone bridge, catalytic bead, metal oxide, infrared, thermal conductivity**, and laser. Readings, capabilities, and reactions can and do vary between the six, and responders' safety depends on that instrument, so they must understand how it works. A common complaint at training is "When we used our meter next to the gas company rep's meter, his was going off, and ours was not reading at all." Or "His reading was twice what our meter stated." There are a couple of reasons for this. To compare readings, you must compare sensor to sensor, and in many cases, the gas company is using a metal oxide sensor (MOS) to look for small leaks, while most responders are using a catalytic bead sensor. There is no question that the MOS is much more sensitive and reacts far more quickly than the catalytic bead sensor. Calibration plays a factor in the readings being double or just different, as the gas company meter is calibrated for natural gas (primarily methane) or propane, and the responders' meter may be calibrated to pentane and may be set to react differently for natural gas. Some meters use a 0.5 correction factor for natural gas, which means that the readings will be off by 50 percent, thereby providing an intentional safety factor. The gas company meter does not have this cushion built in. The other factor is that most gas companies calibrate their meters daily, something very few responders do. Thermal conductivity sensors react to gases other than flammable ones, which can cause confusion. The next section describes how these various sensors work and how the readings may differ.

Wheatstone Bridge Sensor

What used to be the most common sensor, an older version of the wheatstone bridge sensor, was essentially a coiled platinum wire in a heated sensor housing (**FIGURE 5-3**). The wheatstone bridge is also part of the construction of other flammable sensors. When the sample gas passes over the "bridge," it heats up that side of the bridge. Platinum is used because it can catalyze oxidation (combustion) reactions at relatively low temperatures and at low concentrations of flammable vapors. If the air sample contains any concentration of a flammable gas, the platinum filament will get hotter and increase its resistance to an electrical current in a nearly direct proportion to the concentration of any flammable content. The change in resistance is compared to a known constant resistance or a second parallel bridge, and the difference is converted to a meter reading. Wheatstone bridge is old technology but not bad technology, and it has a proven track record. Newer and much better monitors with wheatstone bridge LEL sensors are actually two separate wire coils in the middle of the sensor; both are heated with the sample gas passing over one of the bridges. The sample gas, if flammable, causes the bridge to heat up, resulting in a

FIGURE 5-3 Wheatstone bridge sensor.
© Jones & Bartlett Learning

difference in the electric resistance in between the two bridges. This difference is reported on the readout of the monitor. These new types are a marked improvement over the old and have corrected many of the problems given in the following list. The advantage of the newer monitors is that they allow for correction of ambient temperature, humidity, and other factors that cause instabilities to the meter reading.

Wheatstone bridges may have problems in the following areas:

- Low-oxygen atmospheres. To get an accurate reading, the minimum amount of oxygen necessary is 16 percent. Less than 16 percent oxygen, and the readings will be off and vice versa. If the oxygen is high, then the LEL reading will be high.
- Acute sensor toxins such as lead vapors (from leaded gasoline), sulfur compounds, silicone compounds, and acid gases can corrode or coat the filament, which can cause altered readings.
- Chronic exposure to high levels may saturate the sensor and cause it to be useless for a long period of time until purged and recalibrated.

The old wheatstone bridge LEL sensors (as well as many new ones) do not indicate when they go above the LEL. In most cases, the LEL sensor would indicate 100 percent of the LEL for a very short time and then bounce back to 0 percent, never to rise again, no matter what. The bridge had burned out and would not function. Many of the meters allow the sensor to read up to 100 percent of the LEL, then shut the sensor off, or if so equipped, switch over to the LEL sensor that does percentage by volume methane. This improvement will save many dollars in sensor replacement. The new LEL sensors should last a minimum of 4–5 years, although almost all manufacturers only warranty them for 1–2 years. The wheatstone bridge reads on a scale of 0 percent to 100 percent and starts to indicate at a level of approximately 50 ppm of methane. Pure wheatstone bridge sensors are not in common use anymore but can still be purchased.

Catalytic Bead Sensor

The catalytic bead sensor, also known as a pellistor, is the most common LEL sensor for emergency response and uses a wheatstone bridge as its base component with some new twists. Instead of spirals of wire forming a bridge, the catalytic bead sensor uses a wire with a bowl or a round piece of metal, which we call the bead, in which there is a wheatstone bridge. The wheatstone bridge is made of platinum wire. The metal bead is typically coated with a catalytic material

that helps burn the gas sample off efficiently, at a temperature of 932–1022°F (500–550°C). The application of a catalytic material allows the flammable gases to burn at a lower temperature than their ignition point. Without the catalytic coating, the sensor would have to be heated to the 1652°F (900°C) range, which would degrade the sensor over time. The sensing unit, as shown, has two sensors placed in the same fashion as the wheatstone bridge, one for sampling and the other for reading the change in the sampling bowl (**FIGURES 5-4** and **5-5**). The dual beads compare their electrical activity to provide the meter readings. The dual sensors also have the inherent ability to compensate for ambient temperature, humidity, and atmospheric pressure. The catalytic bead sensor is much more precise than the wheatstone bridge and is less susceptible to breakage. The platinum wire is used as it has good ability to be heated for long periods of time without significant degradation. It can be heated

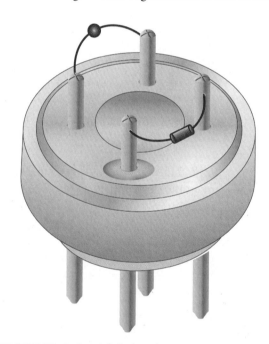

FIGURE 5-4 A catalytic bead sensor.
© Jones & Bartlett Learning

FIGURE 5-5 A catalytic bead sensor.
Courtesy of Industrial Scientific.

to relatively high temperature without impact, but if the temperatures reach the 900°C range, the platinum is impacted and can cause the readings to drift, which is why calibration is important. With normal use, they usually last 4–5 years. The catalytic bead sensor reads at 0–100 percent of the LEL, although some units also read at parts per million levels, usually starting at 50 ppm. One very useful feature on some monitors is the sensor that shuts off at 100 percent of the LEL, which prevents the sensor from becoming saturated or burned out. The sensor is covered by a flame arrestor, known as a sinter, which prevents the flame from escaping the sensor housing. The catalytic bead also begins to read at about 50 ppm. Some newer catalytic bead sensors are being used to read both percent of LEL and parts per million. The ability to switch measuring modes was formerly reserved for the metal oxide sensor. The ability for the catalytic bead sensor to detect small amounts of a flammable material, usually starting at the 50-ppm range, is generally the result of the sample gas being preheated before it enters the catalytic bead chamber. As the gas is heated, it reacts differently in the chamber, and the chamber will burn off the sample, causing a reading on the monitor. The catalytic bead sensor will break down after a few years, but most last up to 5 years. There are several items that can poison the sensor, and some cause temporary effects or may permanently damage the sensor. **TABLE 5-2**

TABLE 5-2 Catalytic Bead Poisons	
Temporary Poisons	**Permanent Poisons**
Hydrogen sulfide	Silanes, silicone, or silicone-based products
Halogenated compounds	Chlorine
Freon	Airborne lead and other heavy metals
Chlorinated compounds such as vinyl chloride, carbon tetrachloride, methyl chloride	Organophosphorus-based materials
Acetylene	Acid gases such as hydrochloric
Sulfur compounds	
Olefins such as propylene, styrene, acrylonitrile	

provides some examples. The sensor can be damaged by silicone or silicone-based compounds, sulfur-based compounds, chlorine, lead, and exposure to some heavy metals. Firefighting foam can damage the sensor as well. Some of these items can poison a sensor with low concentrations and with short exposures. Hydrogen sulfide and halogenated hydrocarbons such as halogen and Freon can temporarily poison a sensor, and if exposed to these gases in high concentrations, the sensor should be allowed to run in fresh air for 24 hours or more to clear the contaminant. Acetylene is another gas that may impact some sensors, and repeated exposure should be avoided. In some cases, when a catalytic bead sensor has been exposed to a poisoning gas, it will fail to the zero point, and unless you bump test the instrument, one would assume that it is functioning fine. There are several ways to reduce the risk to your device. The first is to reduce the amount of time the device is in the damaging environment. Other methods include the use of filters and changing the pump speed to reduce the amount of material that is entering the device. There are occasions and scenarios where you must use your device in a damaging atmosphere as there may be other risk factors present. If you enter an environment that may have multiple hazards, and one or more of them involves damaging materials, perform regular bump tests on the device to make sure it is still functioning. It is important to ask the sales rep what type of sensor the LEL sensing unit is, as there is a difference between the catalytic bead sensor just described and the other sensors discussed in the next section.

Metal Oxide Sensor

The metal oxide sensor (MOS) or broadband sensor has attracted a lot of negative attention. Most people do not understand how the sensor functions, and since it is a very sensitive sensor, it has a great deal of perceived problems. If used and interpreted correctly, this sensor can provide answers to many response questions (**FIGURE 5-6**).

Most of the emergency services' MOS devices use a tin dioxide element, which, in the presence of reducing gases such as propane, methane, and other flammable gases including CO, causes a change in resistance. The coating on the sensor determines its sensitivity and applicability. Tungsten trioxide-coated MOS sensors are more sensitive to hydrogen and nitrogen oxide but do not react well to ammonia. In simple terms, when a sample gas hits the sensor, there is an increase in heat, which causes a change in electrical resistance. The actual reaction that takes place at high heat is the electrical change (conductivity) in the sensor due to the

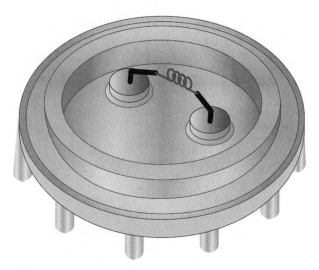

FIGURE 5-6 A metal oxide sensor.
© Jones & Bartlett Learning

absorption/deabsorption of oxygen and the target gas. As the metal oxide coating, which can help make the sensor more specific, reacts with the gas, the sensor can detect low levels of gases. In some cases, the sensor can detect levels in the 1-ppm range. Almost anything that enters the sensor is burned off, and the reaction results in an electrical change, which causes a meter reading. Remember that the wheatstone bridge and catalytic bead sensor react only to flammable gases, nothing else. The MOS picks up dirt, dust, and other particulates; moisture in the air; flammable gases; and even some low levels of vapor from a combustible liquid. If the chemical has sufficient vapor pressure to get into the air, the MOS sensor typically can detect it.

This sensitivity can present a problem to some people, as they do not understand how the sensor works. If anything is on the sensor when the meter is first turned on, it will heat up, trying to burn off the material, causing a reading. Most people are not patient enough to let the meter warm up for at least 5–10 minutes, clean itself, and stabilize. In most cases, this type of sensor requires frequent calibration and requires zeroing each time it is turned on. It can be a very frustrating sensor, but it provides a lot of valuable information about air quality and potential contaminants. It is so sensitive as compared to the other LEL sensors that it can pinpoint tiny leaks in pipes, a valuable benefit in emergency response and industrial use.

LISTEN UP!

The MOS is so sensitive as compared to the other LEL sensors that it can pinpoint tiny leaks in pipes, a valuable benefit in emergency response and industrial use.

Most MOS LEL sensors do not provide a readout of the percentage of the LEL as other LEL sensors do, but they provide an audible warning or a number within a range. If you spill a tablespoon of baby oil on the table and pass a MOS sensor over it, the sensor will give a reading, although no other LEL sensor would pick this material up. If you take a MOS sensor into a room with 5 percent methane, the reading will be within a range for a flammable gas. Some monitors allow the MOS to read in percentages of the LEL, in addition to a general sensing range of 0–50,000 units. The MOS reacts to tiny amounts, which is an outstanding feature of the monitor, as most monitors are not this sensitive. In many cases, hazardous materials teams are called to buildings in which tiny amounts of something in the air are causing problems. The MOS detector is very useful in determining whether something is there and pointing to where the unidentified material is. If the department can only afford one combustible gas sensor, it is not recommended that you purchase a MOS unit. A catalytic bead will give adequate performance at a lesser cost. If a response team is looking to further enhance its capabilities, then a MOS is a great addition to its air monitoring capabilities. The MOS sensor usually can be set to read 0–100 percent of the LEL or on a 0–50,000 unit scale and starts to detect some gases at levels near 3–5 ppm, a great advantage. It is near photoionization detector (PID) levels for some gases and reacts to many things a PID does not.

LISTEN UP!

The MOS picks up moisture in the air, flammable gases, combustible gases, or just about any chemical with enough vapor pressure to get it into the air.

One of the biggest drawbacks of the MOS sensor is its nonlinear output. The catalytic bead sensor is known as a linear sensor. A sensor that has a linear output is one that has readings that can be predicted and duplicated, which is further discussed at the end of this chapter.

Many emerging sensors in the detection world, some of which are nanosensors and are very small, use MOS as their backbone. Some weapons of mass destruction (WMD) detection devices use MOS sensors to help detect very small amounts of chemical warfare agents. For flammable gases, the MOS sensor is coated with a catalytic material, one that burns flammable materials well. To detect other gases, the MOS sensors are coated with other materials. One example in the detection of hydrogen is the coating of a sensor in zinc oxide (ZnO), which is highly reactive to hydrogen. These are coated on nanorods, which are 2–10μ in length. Other coatings include titanium (Ti), nickel (Ni), silver (Ag), gold (Au), platinum (Pt), and palladium (Pd). Carbon nanosensors have shown good detection ability for ammonia, carbon dioxide, alcohol, and nitrogen oxide. These sensors can detect these materials in small quantities and do not require any substantial heating of the sensor.

LISTEN UP!

In many cases, hazardous materials teams are called to buildings in which tiny amounts of something in the air are causing problems. The MOS detector is very useful in determining whether something is there and where it may be coming from.

Thermal Conductivity Sensor

The thermal conductivity (TC) sensor exists in some multigas instruments (**FIGURES 5-7** and **5-8**). As can be implied by the name, thermal relates to the temperature of the sensor or more specifically the temperature impact (conductivity) a gas has on a sensor. The TC sensor has two sensing points, of which a wheatstone bridge is the main component. Both are heated, and one is known as the reference, which is sealed and is not exposed to the sample gas. The gas enters the sensor and crosses the detection sensor, and the gas may cause a temperature decrease. This change in temperature is referenced with the reference unit, and the difference results in a reading. Methane causes the TC sensor to react and cool, creating a reading. Other flammable and some nonflammable gases also cause a temperature drop and therefore a reading. For the TC sensor to detect a gas, it must have a TC rating of less than that of air. The TC sensor can be set to read parts per million or can read in percent by volume up to 100 percent. One advantage is that the TC does not require oxygen to function. A disadvantage with a TC sensor is that it does not detect all flammable gases. Luckily, RKI Instruments provides a device that has both a TC sensor and a catalytic bead sensor so that you get the coverage for all flammable gases with the catalytic bead, and you get the full range of the TC sensor.

In Chapter 4 we discussed an incident at a fast food restaurant where there were some initial challenges in determining the source of the problem. When the hazardous materials crews entered the building and

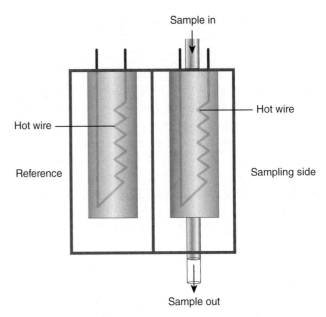

FIGURE 5-7 Thermal conductivity sensor.
© Jones & Bartlett Learning

FIGURE 5-8 An instrument that has both a thermal conductivity (TC) and a catalytic bead (CB) sensor.
© Jones & Bartlett Learning. Photographed by Glenn E. Ellman.

went to the basement, they reported that they were getting oxygen readings of 17.5 percent. This is a drop in oxygen of 3.4 percent, which translates to 170,000 ppm of a contaminant being present. They also reported an LEL reading on their instrument, and it was determined that the LEL sensor was a TC sensor. When it was determined that the contaminant was CO_2, this oxygen drop means that the atmosphere contained 17 percent CO_2. The TC of CO_2 is greater than that of air, and with that volume of CO_2, it caused the TC sensor to indicate.

Infrared Sensor

The infrared sensor has some unique features that further enhance the hazardous materials team's capabilities. Infrared sensors are sometimes referred to as nondispersive infrared absorption (NDIR). When infrared light is passed through certain gases, the gas will absorb some of the light. Light consists of varying wavelengths and is a known quantity. The infrared sensor uses a hot wire to produce a broad range of wavelengths, uses a filter to obtain the desired wavelength, and has a detection device on the other side of the sensor housing (**FIGURES 5-9** and **5-10**). The light emitted from the hot wire is split, one part through the filter and the other to the detection device to be used as a reference source. When a hydrocarbon gas is sent into the sample chamber, the gas molecules absorb some of the infrared light that does not reach the detection device. The amount of infrared light reaching the detection device is compared to the reference source, and if there is a difference, the meter outputs a reading. The readings are also impacted by temperature and atmospheric pressure, and both are compensated for within the sensor housing. This sensor is also very stable and does not require as frequent calibration as the other LEL sensors.

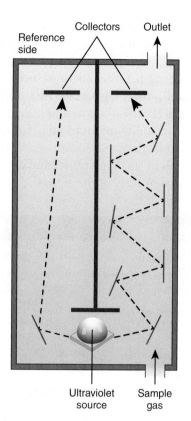

FIGURE 5-9 Infrared sensor.
© Jones & Bartlett Learning

FIGURE 5-10 Infrared sensor.
Courtesy of Christopher Hawley.

LISTEN UP!

Nondispersive infrared (NDIR) sensors shoot the infrared (IR) light without it being dispersed by a prism. A dispersive infrared (DIR) sensor shoots the light through a prism, which allows only a specific wavelength of light to go through the sample. The NDIR shoots all the light through the gas sample. NDIR is used for emergency response devices.

The big advantage of infrared over other LEL sensors is that it does not require oxygen to function; it can read flammables in oxygen-deficient atmospheres. One common situation in which this capability is useful is in ship fires or other situations where a response tactic is to inert the atmosphere. Inerting agents such as nitrogen are commonly used in cargo hold fires or in situations in which materials are reacting in a hold of a ship. When atmospheres are inerted, usually with nitrogen, the goal is to have a 0 percent oxygen atmosphere. The easiest way to determine if this tactic has worked is to monitor for flammable gases, but the low oxygen level presents some stumbling blocks for monitoring. Most infrared meters also measure up to 100 percent by volume. When the levels exceed 100 percent of the LEL, the meter then switches to percent by volume, which means it can read upward from the LEL to 100 percent by volume. The device also is not affected by temperature, nor is it easily poisoned by high exposures, but the sensor may take a while to clear. The sensor also requires a

flow of sample gas to determine the level present, so it may not detect transient gas pockets. The catalytic bead sensor is a direct measurement device for a flammable material; the infrared sensor is not. The infrared sensor measures the absorption of infrared energy at 3.3- to 3.5-micron (μ) range, which is characteristic of a carbon–hydrogen bond. Methane resonates at 3.4μ, which is right in the middle of the detection range. Various filters within the sensor make the sensor more specific. The infrared sensor misses those materials without a carbon–hydrogen bond; some major materials that are missed include hydrogen, acetylene, acrylonitrile, aniline, and carbon disulfide. The lack of this sensor to effectively detect hydrogen and acetylene is a large one. Hazardous materials teams do see acetylene and hydrogen during their responses. Hydrogen is typically detected when dealing with overheated batteries. See Chapter 14 for more information on hydrogen. There has been some discussion in the air monitoring community that to compensate for the fact that the sensor cannot see hydrogen or acetylene the user should rely on the CO sensor. Although true that a CO sensor does have cross sensitivities for both gases, it would not be advisable to rely on the cross sensitivities of a sensor to detect two very dangerous gases. Certainly, if a CO sensor indicates the presences of a gas, one should take advantage of this piece of information and use other devices to assist in keeping responders safe, but one should not rely on a cross sensitivity as his or her first tool into a dangerous situation. There is certainly a need for NDIR sensors in hazardous materials response, but it is but one tool in your detection toolbox. Disadvantages also include cost and many cross sensitivities.

SAFETY TIP

The infrared sensor does not have the ability to detect all flammable gases and cannot detect hydrogen or acetylene.

Laser Sensors

Although not common in emergency response, there are fixed facility laser devices that work in the same fashion as the infrared sensors. It measures the return energy back to the device, and the calibrated gas, if present, will absorb a certain amount of energy. It is only good for the detection of methane, its calibrated gas. The class of lasers varies with the device; some use class 1, 2, or 3 lasers for analysis and guidance. One device sends the readings via Bluetooth to your smartphone, and it measures in parts per million or percent by volume. It can detect 100 ppm methane at a distance of 33 feet (10 m), and the lower detection threshold is 1000 ppm at a distance of 98 feet (30 m). These devices have a quick response time and offer a result in less than a second. Although it is a standoff device, one interesting note is to make sure that you can use this device in flammable atmospheres, as not all units are rated for these types of atmospheres.

LISTEN UP!

The big advantage of infrared over other LEL sensors is that it does not require oxygen to function; it can read flammables in oxygen-deficient atmospheres.

Linearity

When discussing the various LEL sensors, the subject of linearity arises and how it applies to response work. The wheatstone bridge, catalytic bead, and infrared sensor are linear sensors. Although the IR sensor is only linear for its calibrated gas, it uses software to make the other readings linear. By being linear, they can provide estimations of an anticipated concentration. By knowing what the concentration is at one point, you can calculate what the concentration will be when the time is extended. Record one spot on the graph at 0 percent, which is one known point, and then record the next point, which is another known. When you plot these out and extend the line, you can anticipate future concentrations. If we expose a catalytic bead sensor to 25 percent of the lower explosive limit for methane, which is 1.25 percent of the methane by volume, and we calibrate the sensor to that level, the sensor is set to detect methane accurately.

LISTEN UP!

Most infrared meters also measure up to 100 percent by volume in air. When the levels exceed 100 percent of the LEL, the meter then switches to percent by volume in air.

Because the sensor is linear, if we expose it to 50 percent of the LEL for methane, the sensor knows it is seeing 2.5 percent methane by volume in air. If the sensor is exposed to 12.5 percent of the LEL for methane, it knows that the level is 0.625 percent of methane by volume in air. Another way of describing linearity is that the readings are accurate to the actual concentration across some portion of the scale, or the whole scale, before the reading deviates between actual and measured amounts. If you are trying to read a gas other than the one the IR sensor is calibrated for, it will not be linear, and readings will vary. A MOS is not linear, and once the humidity or temperature changes, the sensor is no longer going to read the same as the previous readings. When exposed to a contaminated atmosphere, the sensor is likely to undergo changes, and the amount of contaminant determines how far the sensor moves from the calibrated value. Even in known concentrations where the level does not change, the reported readings will be slightly scattered. A MOS is valuable as it detects tiny amounts of things in the air but is irritating as it is nonlinear. **TABLE 5-3** provides an overview of all the sensors.

TABLE 5-3 Overview of All the Sensors

Sensor	Wheatstone Bridge	Catalytic Bead	Metal Oxide	Infrared	Thermal Conductivity
Detection range	0–100% LEL	0–100% LEL	N/A	0–100% volume	0–100% volume
Response time (T90)	10–30 seconds	10–30 seconds	<1 second	30 seconds	30 seconds with 5-foot hose
Linear?	Linear	Linear	Not linear	Only linear for the calibrated gas	Linear
Accuracy for calibrated gas		±5%	Not calibrated		±5%
Warrantied life/ typical lie	<5 years	2–3/5 years	3 years	1–2/2–3 years	2 years
Temperature range		−40–60°C	32–122°F (0–50°C)	−40–122°F (−40–50°C)	−40–122°F (−40–50°C)
Drift (annual)		5–10%		<5% signal/ month	
Susceptible to poisoning?	Yes	Yes	Yes	No	No, but response time may increase

© Jones & Bartlett Learning

After-Action REVIEW

IN SUMMARY

The ability to detect the fire hazard is important because flammables can cause problems over a large area, and you could be in it. Some flammables are also toxic, so they present a double risk. Flammables and combustibles are the leading category of materials that records show are released in this country, and they are our most frequent responses. Knowing the type of LEL sensor in your instrument is key to solving many air monitor issues. Knowing the various types of sensors that exist makes you an educated consumer and allows you to purchase instruments that best suit your needs. For responders who are going to respond to natural gas (methane) and propane leaks and flammable liquid incidents, the catalytic bead is the best choice for the initial purchase. For hazardous materials teams who have a good complement of catalytic bead sensors, purchasing some of the other technologies is advantageous. Study the advantages and disadvantages, and based on your most likely responses, this can help guide you as to what sensors to purchase next.

KEY TERMS

Catalytic bead sensor The most common type of LEL sensor; uses two heated beads of metal to detect the presence of flammable gases. Also known as a pellistor.

Combustible gas indicator A old name for a device designed to measure the relative flammability of gases and to determine the percent of the lower explosive limit. Also known as an LEL monitor.

Flammable gas indicator (FGI) A more appropriate name for a device that indicates when flammable gases are present. Also known as an LEL monitor.

Infrared sensor A type of LEL sensor that uses infrared light to detect flammable gases.

LEL meter The best name for a meter that is used to detect flammable gases.

LEL sensor A sensor designed to look for flammable gases; can be of four designs: Wheatstone bridge, catalytic bead, metal oxide, and infrared.

Metal oxide sensor (MOS) A form of LEL sensor.

Thermal conductivity sensor A type of LEL sensor.

Wheatstone bridge A form of LEL sensor.

Review Questions

1. A standard LEL sensor reads up to what level?

2. What is the most common type of LEL sensor?

3. What happens to most LEL sensors when they reach 100 percent of the LEL?

4. What type of sensor also indicates levels above the LEL and percentage by volume?

5. Which type of sensor is linear: Catalytic bead or metal oxide?

6. What does a catalytic bead sensor detect?

7. Which LEL sensor does not require oxygen to function?

CHAPTER 6

Toxic Gas Sensors

OUTLINE

- Learning Objectives
- Introduction
- Toxic Gas Sensing Technology
- Carbon Monoxide Incidents
- In Summary
- Key Terms
- Review Questions

LEARNING OBJECTIVES

Upon completion of this chapter, you should be able to:

- Identify the common types of toxic sensors
- Describe the common types of toxic sensor technologies
- Describe common types of carbon monoxide incidents

Hazardous Materials Alarm

Corrosive Atmospheres, Paint, and Salespeople—Corrosive atmospheres, paint, and salespeople are not related, but both stories relate to toxic gas monitoring. Teams responding one evening to a reported drum leak in a warehouse had little additional information. The identity of the leaking material would take some recon. The first responders entered the building to check conditions, taking with them a four-gas air monitor (O_2, LEL, H_2S, and CO), which was a good idea. They reported that they were getting both CO and H_2S readings of about 70 to 90 ppm and were advised by the hazardous materials team to leave the building as quickly as they could. It would be unusual to find both CO and H_2S in the same place at the same time. Also, there were no reported odors associated with the response, and H_2S stinks of rotten eggs. As it turns out, there was a leaking drum emitting corrosive vapors, which was causing the electrochemical sensors to read.

In another incident, an elderly woman called headquarters, and the call was routed to me. An engine company had been to her apartment to check for CO, and for some reason, they did not leave her any details about the incident. I took her name and contact information and said I would call her back. I called dispatch to see who had handled the call, and then I called the engine captain. He said that in an adjacent apartment they had found some open paint cans, and that is what caused the CO. I advised the captain to get back on their engine and return to that address as rapidly as possible. I called the woman back and asked her to kindly step outside and that the fire engine would be right back, but it would be best if she waited outside. I then called the dispatch center to have them send the hazardous materials unit to the incident and to have the officer call me when they were responding. When he called, I told him to find the source of the CO and when they did then educate the engine captain on what causes CO. It is only caused by incomplete combustion, and in this case, the woman's hot water heater was malfunctioning and was emitting 200 ppm CO. Luckily, the captain had neglected to inform the woman about the details, and she was lucky enough to call headquarters.

The salesperson portion of this case study relates to my knowledge of air monitoring. I am not afraid to ask questions, so when a monitor does not do what I think it should, I ask why. We also use many monitors in training situations, and we like to use live chemicals to show how various monitors respond. One brand of monitor, when placed in proximity to a flammable liquid, provided high readings on the CO sensor, which was unusual and caused problems in explaining the reason to the students. The CO sensor was becoming saturated and reacting, causing a reading. None of our other manufacturers' monitors did that, so I took the cover off of a couple and compared the CO sensors. All the other sensors had a filter on top of them, but the reacting monitor did not. I discussed this issue with the manufacturer's sales rep, who defended the product but stated that he would get back to me. The company sent an engineer, who stated that their meter was better than any other and that the other manufacturers' meters were malfunctioning. After discussing the calibration issues and how we were using these meters for training, he still maintained that his were better. When asked how CO is emitted from acetone in a jar, his answer was that this was just training, and it would not happen in real life. We went outside to an area where we conduct chemical training, where there is a recovery system for spilled materials. At the spill pad, we spilled a larger amount of acetone and showed that his meter was reading CO, and none of the others were. He didn't have an answer for me and left pretty quickly. A week later, I got a new pump cover for the meter that had a filter built in it, just like the others. The moral of the story is always try the meter, especially in situations in which you will be using them, and trust the meter; they do not lie.

1. Does a CO sensor only detect CO?

2. What causes CO?

Introduction

This chapter discusses the sensors most commonly used in three, four, or five-gas units (LEL, O_2, Toxic/Toxic/Toxic) (**FIGURE 6-1**). These are often used to measure carbon monoxide (CO), hydrogen sulfide (H_2S), and chlorine (Cl_2). **Toxic gas sensors** are available in a variety of materials, however, and include the capability to detect ammonia, sulfur dioxide, hydrogen chloride, hydrogen cyanide, nitrogen dioxide, and many others. The most common unit sold today by far is a four-gas unit that measures LEL, O_2, CO, and H_2S; its popularity is a direct result of the confined space regulation issued by OSHA. In this regulation, OSHA says that, before the employee enters the space, the internal atmosphere must be tested with a calibrated, direct-reading instrument for oxygen content, for flammable gases and vapors, and for potential toxic air contaminants, in that order. OSHA uses the phrase "potential toxic air contaminants" because the contaminants vary with the situation. It is commonly thought that a four-gas monitor handles all confined space situations, which is not true; the toxic sensor must match the threat.

Some responders use single-gas toxic meters (**FIGURE 6-2**). These instruments have the advantage of being small. They are also relatively inexpensive—an especially nice advantage if they are destroyed by the hazardous atmosphere. (Some atmospheres may be harmful to the sensors in a multigas unit.) The disadvantage is that they detect only one gas; they do not provide the oxygen content or a warning about other potential hazards. They have some advantages in an industrial application, but in an emergency response situation, they must be used in combination with other instruments.

FIGURE 6-1 A multigas unit with CO, HCN, and LEL sensors.
Courtesy of Christopher Hawley.

FIGURE 6-2 Single gas unit with an H_2S sensor.
Courtesy of Industrial Scientific.

Many response teams consider some of the mentioned toxic sensors such as chlorine and ammonia as their fourth or fifth gas. This decision may not be wise economically, as these sensors are costly to calibrate and operate. Typically, hazardous materials teams use colorimetric tubes to monitor for these other gases. Colorimetric tubes cost about $150 per set of 10 tests and have a shelf life of 2 years. Response teams should determine their department's annual cost of doing this colorimetric sampling and compare it to the upkeep of the toxic sensor. The average cost for a CO or H_2S sensor

is about $175, but the cost of a toxic sensor for chlorine or ammonia is $400 to $500. These sensors usually last less than a year; the calibration gas is $300 to $400 and has a 6-month shelf life. Colorimetric sampling is discussed in Chapter 8. The other problem is that if used in more than 20 ppm chlorine, the entire unit could be ruined. For teams that respond to ammonia or chlorine leaks on a regular basis and who have the financial resources for only one unit, a single gas unit is recommended for purchase for that response. Having just a single gas unit limits the potential damage to that unit. Keep in mind that ammonia is a constituent of most heavy-duty refrigeration systems used for food, beverage, and other cold storage and, thus, may be the most-needed monitor. Water treatment systems (drinking or wastewater) would likely have chlorine.

Toxic Gas Sensing Technology

Most toxic sensors are **electrochemical sensors** that function somewhat similarly to oxygen sensors (**FIGURE 6-3**). The electrolyte in the sensor reacts specifically to the target toxic gas. The electrolyte is a corrosive solution, either an acid or a base, but typically a strong inorganic acid. The electrodes are matched to the electrolyte so that, when the target gas enters the sensor, there is a chemical reaction that results in electrical activity, which is read by the electrodes.

SAFETY TIP

Carbon monoxide and hydrogen sulfide are the most common gases found in confined space entries, particularly sewers and manholes.

Three electrodes in the sensor measure the electrical activity in the sensor. The target gas creates a chemical reaction, which results in a change in the electrical activity, which is read by the electrodes.

FIGURE 6-3 Toxic gas sensors.
Courtesy of Industrial Scientific.

The target gas of the sensor depends on the electrolyte, the electrodes, or the diffusion membrane on top of the sensor. The diffusion membrane, another term for filter, is designed to block **interfering gases** but allow the target gas to pass. In most cases, the diffusion membrane does an adequate job of preventing low levels of interfering gases into the sensor, but when hit with large quantities, it can be overwhelmed. A diffusion membrane that has been overwhelmed by an interfering gas is said to have been "saturated" and yields erroneous readings. Unlike the oxygen sensor's lead wool, which degrades into lead oxide and eventually causes the sensor to stop functioning, the third electrode in these sensors regenerates the chemical process.

For some of the problems related to toxic sensors, we must refer to the chemistry of hazardous materials and the periodic table, which plays an important factor when dealing with toxic sensors. The periodic table is set up with families in vertical columns in which all elements in a column have similar characteristics. Using the chlorine sensor as an example, chlorine belongs to the halogen family. The halogen family also has fluorine (gas), bromine (liquid with high vapor pressure), iodine (solid with a slight vapor pressure), and astatine (solid). Each of these materials has similar characteristics, and the sensor reacts to vapors from them similarly. Although the sensor cannot distinguish among these materials, it reacts and provides a reading. Responders who do not know what they are sampling could misinterpret.

An additional complication is that other chemicals can cause interference with the sensor. Interfering gases are listed in **TABLE 6-1** and are also described in the case study Corrosive Atmospheres, Paint, and Salespeople. An example of such an interfering gas for the CO sensor is H_2S. Most CO/H_2S sensors are filtered or manufactured to eliminate the interference, but some are not, so buyer beware. At times, a simple charcoal filter over the sensor makes a great difference in obtaining accurate readings. In some cases, high flammable gas readings bleed over to the CO and H_2S sensors, causing false readings, usually when the LEL sensor is reading about 40 percent or higher. Before you buy a meter, test several units, and expose them to a variety of materials to see how they react. The CO and H_2S sensors in theory should not react to a flammable liquid such as acetone. If one or both toxic sensors does react to the acetone, the sensor(s) are not properly filtered, and a different meter should be considered. When a device is filtered, ensure that you follow the manufacturer's recommendation for changing of the filter. When you expose the device to a volatile organic compound (VOC) such as acetone, and the CO sensor indicates, it is time to change the filter. A filter saturated with alcohol will also cause the sensors to indicate a reading. The toxic sensors and

TABLE 6-1 Interfering Gases for CO and H₂S Sensors

Acetylene	Isopropyl alcohol
Ammonia	Mercaptans
Carbon disulfide	Methyl alcohol
Dimethyl sulfide	Methyl sulfide
Ethyl alcohol	Nitric oxide
Ethyl sulfide	Nitrogen dioxide
Ethylene	Phosphine
Hydrogen	Propane
Hydrogen cyanide	Sulfur dioxide
Hydrogen sulfide	Turpentine
Isobutylene	

FILTERED VS. UNFILTERED CO SENSOR

Contaminant/Level	Without Filter	With Filter
Butane 100 ppm	1 ppm	1 ppm
Acetylene 250 ppm	250 ppm	0
Hydrogen 1000 ppm	420 ppm	360 ppm
Hydrogen 100 ppm	40 ppm	31 ppm

Excerpted from RAE Systems by Honeywell Technical Note-121.

the oxygen sensor have a variety of shelf lives, with many lasting 12–18 months. The amount of time they last depends on the kind of atmospheres to which they are exposed. The sensors have a maximum exposure limit; surpassing this limit kills the sensor. Usually low levels over a period of time do not dramatically change the life of the sensor; high exposures shorten the life of the sensor. When other gases that should not affect a sensor do affect it, its life is shortened.

Other factors impact the oxygen sensors. Low and high temperatures can cause the sensors to react slowly or to yield erroneous readings. The electrolyte can freeze at temperatures lower than 32°F (0°C), and as the temperature approaches this level, the response of the sensor is slowed. Temperatures above 104°F (40°C) cause the electrolyte to become more fluid and lose the ability to react to the target gas. Also, low humidity levels, over time, may affect the sensor and its ability to react to the target gas. High levels of humidity can have an impact on the sensor, and if the sensor is exposed to high humidity for long periods of time, you may see some change in readings. Atmospheric pressure typically does not have any effect on the sensors, but a lack of oxygen does. Low levels of oxygen result in erroneous readings because some oxygen is required for the chemical reaction. As in the flammable sensor, oxygen levels above 16 percent are required for accurate readings.

Some toxic sensors use a metal oxide sensor (MOS), the same type of sensor used for flammable gases. These sensors have the same issues and are not very reliable for toxic gas detection. Electrochemical sensors are linear, and MOS sensors are not linear. The MOS sensors may change their detection levels and response daily. Conditions involving high humidity cause the sensor to provide false readings. Frequent calibration is required because the surface of the sensor undergoes changes that alter readings, and the MOS sensors tend to "forget" what they are designed to detect. If they do not see the calibration gas or the actual target gas, they develop a dead spot and do not see some levels of the target gas.

LISTEN UP!

The CO and H₂S sensors should not react to a flammable liquid such as acetone. If one or both toxic sensors do react to the acetone, the sensors are not properly filtered.

Smart sensors have some promise in the future as the technology advances (**FIGURE 6-4**). The thought

FIGURE 6-4 A set of smart sensors, which can be added to an instrument prior to a response.
Courtesy of Christopher Hawley.

process behind smart sensors is that departments keep a variety of toxic sensors in their inventory, and when they need to sample a particular gas, they just plug in the smart sensor. The smart sensor has a computer chip imbedded in the sensor that tracks calibration and other gas specific information the meter requires. Although one would think that this system would be perfect for emergency response, some versions still needs some improvement. In an optimal situation, the sensors are calibrated and then placed into the storage block, which maintains the sensor. You can place this device into the instrument without calibration. Some manufacturers provide a kit, and each time you swap out the sensor and replace it, it requires calibration, and you need calibration gas for each sensor. The idea behind smart sensors is a good one, but responders should understand the requirements of swapping the sensors and calibration.

Sensor Information

Compare the varying toxic sensors, and when buying a device or a sensor, ensure that you understand the capabilities and the deficiencies of the sensor.

Sensor	Range (ppm)	Expected Life	Response (T^{90})	Cross Sensitivities
Chlorine	0–100	2 years	<60 secs	Yes
Ammonia	0–100	1 year	<60 secs	Yes
Hydrogen Cyanide	0–50	2 years	<200 secs	Yes
Hydrogen sulfide	0–100	2 years	25–40 secs	Yes
Phosphine	0–5	2 years	<60 secs	Yes

Abstracted from CiTicell Specifications for each sensor.

Carbon Monoxide Incidents

Responses to carbon monoxide detector alarms are routine responses for the fire service. In 1995, the city of Chicago ran several thousand carbon monoxide detector alarms in one day due to a temperature inversion, which kept the smog, pollution, and carbon monoxide (CO) at a low elevation within the city. CO is a big issue for the Chicago Fire Department, and in 2012, they placed single-gas CO devices on their emergency medical services (EMS) response units. As carbon monoxide is colorless, odorless, and very toxic, it is important that first responders understand the characteristics of carbon monoxide and how home detectors work. As with other chemicals, CO can be an acute or chronic toxicity hazard. On average, more than 3000 people a year are killed due to CO poisoning. It is only acutely toxic at levels greater than 100 ppm. At levels less than 100 ppm, the hazard comes from chronic exposure. Chronic exposure to CO can be hazardous because many people are exposed to CO in the home. The amount of time people spend in their homes constitutes considerable exposure. One warning, however; CO is often thought of as a wintertime killer, which is far from the truth. CO can be found at any point during the year. Hot water heaters, generators, and grilling are summertime sources of CO that can be found in homes.

Carbon monoxide can only be detected by a specialized detector. In extremely high concentrations, it can be explosive. Exposure to CO causes flu-like symptoms—headache, nausea, dizziness, confusion, and irritability. Depending on the length of exposure, high levels can cause vomiting, chest pain, shortness of breath, loss of consciousness, brain damage, and death.

SAFETY TIP

CO poisoning signs and symptoms may be delayed for 24–72 hours. Levels over 100 ppm are extremely dangerous, and the residents should be medically evaluated. Fatalities have occurred in the developing fetus at 9 ppm.

When exposed to levels of CO in a house, the residents may not exhibit any signs or symptoms. If CO is detected, they should seek medical attention because CO poisoning signs and symptoms may be delayed for 24 to 72 hours. Levels over 100 ppm are extremely dangerous, and the residents should be medically evaluated. Fatalities have occurred in the developing fetus at 9 ppm. Monitoring with a CO detector is essential to determine the possible exposure to CO. People who may only show minor effects of CO poisoning and who would normally be transported to the closest hospital need to have their residences monitored. If high levels are found using a monitor, then the **hyperbaric chamber** may be the best treatment and should not be delayed. Many times a pulse oximeter (oxygen saturation monitor) is used incorrectly to determine the O_2 level in a patient. Patients who have been exposed to CO can cause a pulse oximeter to read 100 percent, as the monitor reads the oxygen molecule in CO as being O_2. The elderly, children, or women who are pregnant

HOME CO SENSOR TYPES

Biomimetic is a gel-like material designed to operate in the same fashion as the human body when exposed to CO. This sensor is prone to false alarms, as it can never reset itself, unless it is placed in an environment free of CO, which in most homes is impossible. The sensor may need 24–48 hours to clear itself after an exposure to CO. The actual concentration of CO at the time the detector sounds may be low, but the exposure may have been enough to send the detector over the alarm threshold. If responding to an incident in which one of these detectors has activated, it should be placed in a CO-free environment until it clears. In the meantime, residents should rely on an alternate CO monitor. They should not be unprotected. The metal oxide detector is similar to sensors used in combustible gas detectors, but it is designed to read carbon monoxide. How successful this design is in only reading CO is subject to debate. Although the metal oxide sensor is superior to the biomimetic sensor, it has some cross sensitivities and will react to other gases. Although it is hoped that the responders would be using a multiple-gas detection device to check a home, it is possible for this type of sensor to alarm for propane. Responders using only a CO instrument may find themselves walking into a flammable atmosphere. Metal oxide detectors can usually be identified by the use of a power cord, as the sensor requires a lot of energy. In most cases, these sensors provide a digital readout. Some of these devices have both a metal oxide sensor and an electrochemical sensor (**FIGURE 6-5**). Once activated, this sensor needs some time to clear itself, usually less than 24 hours. An electrochemical sensor, which may also be referred to as instant detection and response (IDR), is the same type of electrochemical sensor that is in a multiple-gas instrument. It has a sensor housing with two charged poles in a chemical slurry. When CO goes across the sensor, it causes a chemical reaction that changes the resistance within the housing. If the amount is high enough, it will cause an alarm.

This sensor provides an instant reading of CO and does not require a CO buildup to activate. It has an internal mechanism that checks the sensor to make sure it is functioning, which is a unique feature. Out of the three types of residential detectors, based on sensor technology, the electrochemical monitor would provide the best sensing capability.

FIGURE 6-5 This instrument detects both CO and flammable gases. Any number of flammable materials can activate it.
Courtesy of Christopher Hawley.

are especially susceptible to CO and may have had a serious exposure without showing any effects.

If the first-arriving units do not have a CO monitor and victims may be remaining in the residence, personnel are to don functioning SCBA when searching the residence. After determining no victims are present, crews are to ensure that the house is closed up and then wait outside for a CO monitor. Do not enter an area with an activated CO detector without the use of self-contained breathing apparatus (SCBA). If crews find levels that exceed 35 ppm according to the monitor, they should use SCBA to continue the investigation. Crews should be suspicious when responding to reports of an unconscious person or of "several persons down." They should not enter an area without SCBA if it is possible that CO (or other toxic gases) may be present. An air monitor will ensure responder safety from the gases for which it samples.

It is possible that people may be found unconscious due to a natural gas leak, but this is very unlikely. Natural gas, which has a distinctive odorant added to it, is nontoxic and only asphyxiates a person by pushing oxygen out of an area. The only sign of this exposure is unconsciousness or death; any of the flu-like symptoms are due to CO poisoning. If the level of natural gas is high enough to cause unconsciousness, then a very severe explosion hazard is present, and an explosion would be imminent.

It is possible that standard fire department air monitors may not pick up any CO because the CO detectors purchased for homes are made to detect small amounts of CO over a long period of time, and fire department detectors are instant reading and only pick up 1 ppm or more. The fact that responders may not pick up any readings does not mean the residents' detector is defective. Some factors that account for this discrepancy are low amounts of CO or a momentary high level that activated the alarm but dissipated prior to responders' arrival. The amount of time the residence is open dramatically affects readings. Crews are reminded to keep the residence as closed as possible so that the air monitor has a chance to monitor the level of CO. As a reminder, any time you respond to unidentified odors or sick building calls, it is important to remove any people from the building and keep it closed. As the amounts of toxic gases in sick building incidents are usually small, keeping them contained is very important. You cannot treat patients of toxic gas exposure unless you have a clue as to the source. For the patients' long-term health and your own, you need to obtain quick, reliable gas samples.

In some parts of the country, CO detectors are required just as smoke detectors are. Performance of these detectors varies from brand to brand. The three basic sensing technologies are biomimetic, metal oxide, and electrochemical, each with advantages and disadvantages. Location, weather conditions, and the type of sensor determine the types of readings that can be expected from a detector.

Common sources of CO include:

- Furnaces (oil and gas)
- Hot water heaters (oil and gas)
- Fireplaces (wood, coal, and gas)
- Kerosene heaters (or other fueled heaters)
- Gasoline engines running inside basements or garages
- Barbecue grills burning near the residence (garage or porch)
- Faulty flues or exhaust pipes

After-Action REVIEW

IN SUMMARY

The use of toxic sensors partially protects against the toxic hazard, as the toxic sensors are specific to a few gases. Responding to CO events is fairly commonplace and is a response that you should be competent in handling. The response to CO is the most common event in which monitoring is used in combination with patient rescue and treatment. When making toxic gas sensor purchase decisions, keep in mind the costs of operating a meter, and calculate how often the electronic meter would be used.

KEY TERMS

Biomimetic A type of CO sensor used in home detectors.

Electrochemical sensor A sensor that has a chemical gel substance that reacts to the intended gas and provides a reading on the monitor.

Hyperbaric chamber A pressurized chamber that provides large amounts of oxygen to treat inhalation injuries, diving injuries, and other medical conditions.

Interfering gases Picked up by the sensor but not intended to be read by the sensor.

Smart sensor A sensor that has a computer chip on it that allows the switching of a variety of sensors within an instrument.

Toxic gas sensor Device for the detection of toxic gases. Common toxic gas sensors are for CO, H_2S, ammonia, and chlorine.

Review Questions

1. What are the two common toxic gas sensors in a multiple-gas instrument?

2. What types of sensors are used for the detection of toxic gases?

3. Are carbon monoxide sensors specific only to carbon monoxide?

4. Do smart sensors have to be calibrated?

5. What is the danger level for a pregnant woman who has been exposed to CO?

6. Which CO sensor type mimics the human response to CO?

7. What is the issue with the MOS sensor being used for CO detection?

CHAPTER 7

Ionizing Detection Units

OUTLINE

LEARNING OBJECTIVES

Upon completion of this chapter, you should be able to:

- Describe the ionization process
- Describe the various types of ionization detection devices
- Identify the uses for ionization detection devices
- Compare the various types of ionization detection types and their use in emergency response

Hazardous Materials Alarm

The hazardous materials unit was called for mutual aid to an adjacent county to assist with a fire in an apartment building that had been burning for some time but had no known hazardous materials. What was known was that two firefighters were in critical condition at shock trauma, one in respiratory arrest. They were being sent to the hyperbaric chamber for treatment of some unknown respiratory problem. When we arrived, it was determined that about 30 other firefighters were experiencing some problems as well. The fire was still burning, but there was no report of any materials that had been found that may have been the source. Some air monitoring on the scene was done without much success. On a whim, we decided to check the air bottle of one of the downed firefighters. When the photoionization detector (PID) was put into the air stream, it provided a hit. We set up some apparatus to test the air in the self-contained breathing apparatus (SCBA) bottles. We found that the air in the SCBA was providing readings on the PID and that in the colorimetric testing we found some questionable contaminants and carbon monoxide in excess of 70 ppm. These problems were found in many of the bottles we had confiscated. At 3 a.m., we contacted our local lab, and with split samples, we headed to the lab. We would drop the other samples off at the state lab in the morning. That morning the lab came back with the test results from a **gas chromatograph/mass spectrometer** and reported that the air had components of gasoline in it. We identified the source as a compressor unit that had compressed on the scene; the engine was not running well, and there were some concerns with the location of the air intake hose. The PID followed up by colorimetrics was essential to solving this problem. The firefighters recovered and were released a few days later.

1. Normal air does not contain toxic vapors; how does one detect toxic vapors?

2. What is the dangerous level of any vapor called?

Introduction

Ionization is a common sensing technology and is used in a variety of detection devices. This chapter focuses on ionizing devices used for hazardous materials response, while Chapter 11 focuses on ionizing devices used for the detection of potential weapons of mass destruction (WMD) materials. The most common detector of general toxic risks is the **photoionization detector (PID)**, which may be a separate detection device or incorporated into a multigas detection device with other sensors. The **flame ionization detector (FID)** is another detector with the ability to detect potentially toxic materials in the air. Both devices use a form of ionization, which is the creation of charged ions that can be read by a sensor. The use of an ionizing detection device is common in the scientific community. There are several methods for providing ionizing energy through ionization. First and most common in the field is PID, which uses an ultraviolet lamp, while the FID uses a hydrogen-fed flame. Another method of ionization is the use of **corona discharge**, which will be described later. Some devices may use a radiation source to perform the ionization.

IONIZATION

It is important to understand the process of ionization as it applies in this chapter as well as Chapter 11. Gases and vapors are comprised of various atoms, which under normal conditions do not carry a charge and are neutral. It is possible during some conditions that an atom could gain or lose electrons, which would change the electrical makeup of the atom. When they become charged, they are known as ions and carry an electrical charge, either positive or negative. Ionization is possible when electrons are lost or gained.

Ionizing devices are set up in two methods, one that quantifies the amount of material present based in comparison to the calibration gas and/or one that identifies the material. Both require a form of ionization, while the method that identifies adds software and a method to analyze the results of the ionization. When a sample gas enters a device, it is ionized, and it enters the sensing chamber where the positive or negative charged ions are detected. The term "ions" means an atom has a charge and is the key to how ionization identifies a material. A simple ionization device (PID) merely detects the electrical change within the sensor housing, and more complex systems have a sensing tube and track the time it takes the ions to travel down the tube, measuring the activity at the end of the tube. Ions have a unique travel time and, in combination with the identification of the electrical charge, help identify the molecule. For more explanation on how ionization works with devices that identify materials, see Chapter 11.

The method of ionization may vary, but the end result is the same. The resulting change in electrical activity is measured against that of a known gas, which is the calibration gas. The PID, FID, and the corona discharge sensors use ionization to detect a variety of gases, but the mechanism to complete that ionization differs. As can be seen in **TABLE 7-1**, the lower explosive limit (LEL) sensor does not detect at a low enough level to protect responders against toxic risks. For hazardous materials teams, a PID is an essential device, one whose value cannot be overstated. Departments without a PID are taking a serious risk, as they do not have an easy method of detecting common potentially toxic materials.

> ## LISTEN UP!
>
> The PID, FID, and the corona discharge sensors use ionization to detect a variety of gases. These detectors identify the potential toxic risk that chemicals present. The ion mobility spectrometry (IMS) device is the only one that can potentially identify the material.

Photoionization Detectors

Sometimes referred to as a total vapor survey instrument, a PID can detect organic and some inorganic gases, including ammonia, arsine, phosphine, hydrogen sulfide, bromine, and iodine. Because of its ability to detect a wide variety of gases in small amounts, it is an essential tool of response teams. The PID does not indicate what materials are present; it just identifies that something is in the air. Some units are stand-alone units or can be included in a multigas unit (**FIGURES 7-1** and **7-2**). Used as a general survey instrument, the PID can alert responders to potential areas of concern and possible leaks/contamination. The original PIDs were designed for the petroleum refining and storage industry and are widely used during underground storage tank (UST) removals. Because of their sensitive nature, they can detect small amounts of hydrocarbons in the soil. Odor complaints (sick building) calls are on the increase, and the PID is a valuable tool in identifying possible hot spots within the building. PIDs even have a valuable use in possible terrorism events because all the chemical agents can be detected by a PID.

> ## LISTEN UP!
>
> The PID does not indicate what materials are present; it just identifies that something is in the air.

TABLE 7-1 Toxicity Versus Fire Risk

Known Information	Values
Actual spilled chemical	Phenol
LEL	1.8%
OSHA PEL	5 ppm
NIOSH REL	5 ppm
IDLH	250 ppm

Reading from Catalytic Bead LEL Sensor Calibrated to Phenol (%)	Parts Per Million in Air Equivalent
100	18,000
50	9000
25	4500
10 (meter alarms)	1800
3	540
m	180
0.8 (meter would show 0)	141

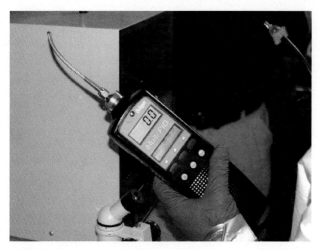

FIGURE 7-1 A stand-alone PID.
Courtesy of Christopher Hawley.

FIGURE 7-2 A multigas PID.
© Jones & Bartlett Learning. Photographed by Glenn E. Ellman.

FIGURE 7-3 A ppb PID.
© Jones & Bartlett Learning. Photographed by Glenn E. Ellman.

TABLE 7-2 Conversion from Percentage of Volume to ppm	
Percentage of Volume	Equivalent ppm
1	10,000
2	20,000
5	50,000
10	100, 000
20	200,000
30	300,000
40	400,000
50	500,000
60	600,000
70	700,000
80	800,000
90	900,000
100	1,000,000

© Jones & Bartlett Learning

The biggest advantage that a PID has is its sensitivity, as most standard devices start to read at 0.1 ppm. Most PID monitors can read up to 2000 ppm, and several monitors read to 10,000 ppm. This reading in parts per million is for the calibrated gas, which is usually isobutylene. RAE Systems by Honeywell has a PID that reads into the parts per billion (ppb), a very sensitive instrument (**FIGURE 7-3**).

When the gas is unknown, we call the reading meter units. A reading of 135 on a PID that reads 0.1–2000 ppm for an unidentified gas would be called 135 meter units, which means the meter moved 135 units out of a scale of 2000. Chemicals with a permissible exposure limit (PEL) or threshold limit value (TLV) of less than 500 ppm are considered toxic, so the PID is useful in identifying these levels. The problem with using the LEL sensors, except for the metal oxide sensor (MOS), is the fact that they require high levels to begin to read. Except for the MOS, LEL sensors do not begin reading unless the levels exceed 50 ppm. We can convert the readings found on an LEL monitor to parts per million by using the formula found in **TABLE 7-2**. Some conversions are provided in Table 6-1.

It is easy to see that one could be in significant trouble in regard to toxicity when an LEL sensor would not even indicate a problem. The PID is used to look for small concentrations in air, and LEL sensors are for the larger problems, most specifically the fire risk.

TABLE 7-3 presents another example as to how it is easy to be fooled into thinking the atmosphere is safe. Suppose that you are responding to a reported odor of gas in an apartment building. Upon arrival you take your four-gas air monitor into the building. When responding to a reported gas odor, at what point do you don your facepiece? What meter reading (LEL sensor) would make you put your facepiece on? Do citizens always know natural gas from other chemical odors? Think about these answers before you read the table.

The table shows that at 0.8 percent of the LEL there are 141 ppm of phenol in the air. The LEL sensor at 0.8 percent would be reading 0, as it requires at least 1 percent to indicate. At 1.6 percent, the meter would read 1, as it only reads whole numbers. If you just used the LEL sensor, you could be in an environment that is more than half the immediately dangerous to life or health (IDLH) without the monitor indicating any level. Many responders use 10 percent or 25 percent of the

LEL in which to put their facepieces on or to take action, which would be 1688 and 4500 ppm, respectively. At these levels, there is a severe toxic risk to an unprotected responder. However, a PID with an 11.7 electron volt (eV) lamp would pick up this material at 0.1 ppm, a level that would offer far greater safety.

SAFETY TIP

Departments without a PID are taking a serious risk, as they do not have an easy method of detecting common potentially toxic materials.

The PID uses an ultraviolet (UV) lamp to ionize contaminants in the air. The gas being sampled is composed of various molecules, and each molecule is made up of a variety of atoms, and each atom is made up of neutrons (neutral), electrons (negative), and protons (positive). Ionization occurs when an electron is removed or added, resulting in a charged particle (an ion) being formed. The sensor detects the energy created with the electrons and the positively charged ion. Because the sensor is calibrated with known quantities of gas, it can equate the change in electrical activity in the sensor to a reading. Once the gas is read by the sensors at the end of the sampling tube, the gas regains an electron and becomes whole again. A stand-alone PID is a nondestructive test, and one can collect a sample from the discharge end of the PID. The lamps for the photoionization detectors have varying strengths, just as regular light bulbs have varying wattages. The PID lamps come in varying strengths and materials, the most common being a krypton-filled lamp with a magnesium fluoride window (**FIGURE 7-4**). The strength of the lamp is 10.6 eV, the most common lamp sold to emergency responders. The lamp strength is the indication of the materials the PID will detect, as shown in **TABLE 7-4**. To be read by a PID, the vapor or gas to

TABLE 7-3 Toxicity Versus Fire Risk	
Known Information	**Values**
Actual spilled chemical	Phenol
LEL	1.8%
OSHA PEL	5 ppm
NIOSH REL	5 ppm
IDLH	250 ppm
Reading from Catalytic Bead LEL Sensor Calibrated to Phenol (%)	**Parts Per Million in Air Equivalent**
100	18,000
50	9000
25	4500
10 meter alarms	1800
3	540
1	180
0.8 (meter would show 0)	141

FIGURE 7-4 PID 10.6 eV ultraviolet lamps.
Courtesy of Christopher Hawley.

TABLE 7-4 Ionization Potential of Common Materials

Chemical	Ionization Potential (eV)	Detected by 10.6 eV Lamp	Detected by a 11.7 eV Lamp
Acetone	9.69	X	X
Acetylene	11.4		X
Ammonia	10.18	X	X
Ethyl ether	9.53	X	X
Formaldehyde	10.88		X
Freon 112	11.30		X
Hydrazine	8.93	X	X
Hydrogen cyanide	13.60		
Hydrogen fluoride	15		
Hydrogen peroxide	10.54	X	X
Hydrogen sulfide	10.46	X	X
Methane	13.0		
Methyl alcohol	10.84		X
Methyl ethyl ketone	9.54	X	X
Nitrous oxide	12.89		
Phenol	8.5	X	X
Phosgene	11.55		X
Propane	11.07		X
Sulfur dioxide	12.30		
Toluene	8.82	X	X
Triethylamine	7.5	X	X
Xylene	8.56	X	X

be sampled must be able to be ionized, which is called **ionization potential (IP)**. The unit of measurement of an IP is electron volts (eV). There are various types of UV lamps available; the most common are a 9.8, 10.2, 10.6, and 11.7 eV. To read a vapor or a gas with a PID, the gas must have an IP less than the eV rating on the lamp. The lamp strength varies according to the gas inside the lamp and the material used to construct the lamp window. The 10.6 eV lamp is the most stable lamp and lasts for more than a year. The 11.7 eV lamp has a window constructed of lithium fluoride lamps and typically lasts 6 months. RAE Systems by Honeywell provides an 11.7 eV lamp in a sealed container. If you need the 11.7 eV lamp, take the 10.6 eV lamp out of the PID, unseal the 11.7 eV lamp, and install it in the PID. At that point, the active life span clock begins. The 11.7 eV lamp is affected by water moisture and starts to degrade once in the open air.

LISTEN UP!

The lamp strength is the indication of the materials the PID will detect.

For example, to sample benzene, which has an IP of 9.2 eV, we must use a lamp of at least 9.2 eV or above. The standard 10.6 eV lamp would be acceptable to read benzene. In most cases, gases that have IPs above the lamp strength cannot be read, although there is some carryover. Some common IPs are provided in Table 6-3. Vapors in excess of the lamp strength will read in some fashion, but it is nowhere near an accurate reading. Keep in mind that the PID is calibrated to a specific material, usually using isobutylene as the calibrating gas, so correction factors apply here as they did with LEL sensors. The LEL sensors have low response factors, but the PIDs have some large numbers for correction factors. Some examples are shown in **TABLE 7-5**. PIDs can pick up readings from toxic substances but also can detect baby oil, motor oil, gasoline, and many other hydrocarbons. Many liquid pesticides are 0.5 percent to 50 percent solution mixed with xylene, trimethyl benzene, and emulsifiers that are easily detected by the PID. Most drum dumps spill waste oil, fuel oil, and the like, so the PID is a valuable resource in protecting responders and the public from toxic materials.

One confusing issue with the PIDs is what is considered to be a toxic reading for unidentified gases when using a PID. Many agencies use the general rule of thumb for an occupancy that has chemicals in use that a reading of less than 50 ppm is acceptable. There is no easy answer when trying to determine what constitutes a toxic environment when using a PID, as many other

TABLE 7-5 Example Correction Factors for PID

Chemical	9.8 eV Lamp CF	10.6 eV Lamp CF	11.7 eV Lamp CF
Ammonia	NR	10.9	5.7
Ethanol	--	9.6	3.1
Ethylene oxide	--	13	3.5
Ethyl mercaptan	0.60	0.56	--
Phosphine	28	3.9	1.1
Propylene glycol	18	4.2	1.6

Excerpted from RAE Systems by Honeywell TN -106.

factors come into play in answering that question. The two primary considerations are location or occupancy and biological indicators. There are certain occupancies, such as print shops, gas stations, paint, and auto parts stores, to name a few, in which a PID will read the vapors in the air. The determination of whether that atmosphere is toxic is based on the predominant material in the air and what that chemical's PEL is. It can be anticipated that in a paint store, where throughout the day containers have been opened and spills are likely to have occurred, vapors may escape their containers. This mixture in the air would be considered normal and probably would not exceed the PEL. You could anticipate readings of 20–300 meter units in this type of occupancy. However, in a home, other than in the garage or the basement, there should not be any toxic vapors in the air, so the PID should read close to the background level. If the PID is used in the garage, it could be anticipated that some readings might be obtained in the area where chemicals may be stored, but you would have to be fairly close to the containers, and the readings should be low. As another example, in a bedroom nothing should cause the meter to go above background. If you do find readings, then there is a problem in that house that needs to be corrected. That does not mean that the problem is toxic or life threatening, only that there are vapors in the air that should not be there and should be investigated to determine if they are hazardous.

The biological indicators (humans) are also crucial to an investigation. If the occupants of the building have

any signs or symptoms and you are reasonably certain that they are real, then any reading above background on a PID should be considered. Such a reading means that there is something in the air in tiny amounts that is causing people problems. Encountering dead people and getting a PID reading of 1 meter unit means that something extremely toxic is in the air, in tiny amounts, and at those levels has killed some people. The best way to really learn how to use a PID and to be able to interpret any readings is to use the PID all the time. When doing inspections or facility tours, take the PID with you. Use the meter in "normal" situations to learn what sets it off and what anticipated readings you may find.

LISTEN UP!

The best way to really learn how to use a PID and to be able to interpret any readings is to use the PID all the time.

Following is a list of some problems with PIDs:

- The biggest factor in the use of a PID is that humidity plays a role in the reading. Some manufacturers have automatic humidity adjustments for their instruments. Some manufacturers provide correction factors for humidity, and other instruments electronically correct for humidity. Humidity can affect PIDs in two ways: First, if a PID has a dirty lamp and a dirty sensing area, the water vapor may create a short, which causes a meter reading. The second and more common effect is the quenching problem. If there is humidity around the lamp, much as fog absorbs the energy of headlights, the meter readings for the chemical of interest are depressed. For all devices, it is best to calibrate your device in conditions as close as you can to the conditions you will be using the device in. This is important with a PID, depending on the calibration gas mixture.
- The lamps are affected by dirt and dust and require cleaning. Environments that have diesel exhaust and other particulate matter such as mown grass may affect the meter. Saltwater or hard water environments may affect the lamp as well. Some devices require hands-on cleaning, while others have a lamp-cleaning mode. The lamp-cleaning mode may require the device to be docked and will turn the lamp on and "cook off" the contaminants, much like a self-cleaning oven.
- Some chemicals such as many corrosives will damage the electronics and possibly the lamp. Phosphine is known to coat the lamp, reducing its effectiveness. Short exposures and frequent lamp

cleaning are recommended when using the PID with phosphine.

- Higher levels of methane (natural gas, swamp gas, landfill gas) may suppress some of the ionization potential of the lamp. Use an LEL monitor to read the LEL because a PID does not read methane (IP of 13.0).

- The PID cannot separate out gas mixtures, and since it does not identify, you are not able to determine the mixture. Using various strength lamps can help separate out a gas mixture, but to do so takes extensive time and thought.

- With some PIDs you may need at least 10 percent oxygen in the air for the PID to function. The RAE Systems by Honeywell PID, for an example, does not require any oxygen to function and has been tested in oxygen-deficient atmospheres.

- Materials that have a high molecular weight or are combustible as opposed to be being flammable are a challenge to PIDs depending on the configuration of the device. For multisensing units, there will be some loss of material, or the material may not travel to the PID portion of the device. The readings may be low, slow, or missing. Some examples include diesel fuel and antifreeze (ethylene glycol). Stand-alone PIDs will typically not have this issue. Materials that react may react prior to reaching the PID portion of the device as well. Iodine and hydrazine are two examples of these materials that will not likely be read accurately by a multisensing device.

Flame Ionization Detectors

Flame ionizations are another method of ionizing a gas sample. These devices are also on the front end of some **gas chromatographs**, which are discussed later in this chapter. This class of instruments used to be referred to primarily as organic vapor analyzers (OVA), but FID is the current term used to describe them. The FID works on a principle like the PID but has some capabilities beyond a PID. Instead of using UV light to ionize any gases that may be present, the FID uses a hydrogen flame to provide the ionization process (**FIGURE 7-5**). Differing from the PID the FID is a destructive test, which destroys the gas sample. The use of hydrogen to provide the flame can be considered a disadvantage because the FID only detects organic materials in the air, whereas the PID detects both organic and some inorganic vapors. The FID does not detect hydrogen sulfide, formaldehyde, carbon disulfide, ethylamine, chloroform, or carbon tetrachloride. The FID also does

FIGURE 7-5 FID.
Image Courtesy of The Thermo Scientific™ TVA2020 Toxic Vapor analyzer.

not detect the components of air. The capability of the FID to detect hydrocarbons that are halogenated, hydroxyl, or amino component hydrocarbons is diminished. The more flammable a material, the easier it is for the FID to detect a material. Using both tools is the start of the detection process. If the PID detects the material and the FID does not, then the material is inorganic. With regard to methane, the PID will not detect methane, while the FID detects methane very well, and in some cases, it is calibrated with methane. If you are detecting a gas other than methane, there are correction factors that must be applied, and some examples are shown in **TABLE 7-6**. If your LEL sensor is indicating the presence of a flammable gas, and the FID is indicating as well, the suspect material could be methane, and naturally occurring methane does not have any odor. If they both detect the material, then it is organic. Because it can read so many varied gases, it generally gives

TABLE 7-6 Example Correction Factors for a FID				
Chemical	100 ppm	500 ppm	5000 ppm	10,000 ppm
Acetone	3.125	2.732	1.664	0.920
Acrylic acid	0.476	0.439	0.356	0.344
Ethylene oxide	3.04	2.83	2.33	2.06
Methanol	23.81	N/A	22.67	22.17

Excerpted from Inficon Technical Note Response Factors for FID Operation.

some type of reading, which can be misinterpreted. PID sensors will indicate flammable materials, such as hydrocarbons, but their effectiveness depends on which family the chemical belongs. A PID will detect materials from the ketones family very well, acetone being an example. It will detect but will have varied effectiveness with alcohols, as a 10.6 eV lamp does not detect methanol. They will detect them depending in the ionization potential. The FID's effectiveness depends on the carbon molecules. The heavier the molecule with more carbon atoms is slow to be detected. If any of the carbon atoms are replaced by a chlorine atom, this challenges the FID. As the FID is a flame-producing device, it potentially could be an ignition source, so to prevent explosions, the sensor uses a flame-arresting filter that prevents the flame from escaping the sensor. It acts much like the flame arrestor used in the LEL sensor, but as with the LEL sensor, heavier gases will not be picked up as well. Taking a PID up to a closed sample jar that contains a flammable liquid will generally not result in any reading above background. When using a FID, though, vapors escaping the closed container can be detected. When a PID ionizes a gas sample, it is a one-time chance for a successful ionization; the FID, however, continues to try to ionize the sample, allowing for even greater results. Because the FID uses a flame to ionize the gas, and the resulting release of carbon atoms is a repeatable event, the FID is a more accurate device than the PID. The FID can read methane gas, while the PID does not read methane, and high levels of methane will depress any PID readings for other materials that may be present.

The FID reads methane very well, down to 0.5 ppm. The FID reads methane, while the PID does not, because the FID can read materials with an ionization potential of 15.4 eV. The highest a PID reads is 11.7 eV, and that is with the special 11.7 eV lamp. The FID has an expanded ability to detect chemicals with a higher ionization potential. As with other instruments, the FID needs adequate oxygen to work, and high wind conditions may cause flame problems. At least 14 percent oxygen is required for accurate readings. Some units require at least 17 percent oxygen to ignite the hydrogen flame during startup. The hydrogen cylinder on the unit may need to be refilled during long duration sampling, and doing so may be cumbersome. The FID does very well at detecting a small amount of toxic material in the air, as well as on the high-end readings that auto scale up to 50,000 ppm. When dealing with concentrated vapors, such as those in the headspace of a drum, there is a tendency for the FID to flame out due to the oxygen-deficient atmosphere. As was suggested with the PID, the best way to get familiar with this device is to use it routinely during nonemergency situations to see what it reacts to and what it does not.

Thermo Scientific produces the TVA2020, a device that combines the PID and the FID within one detection device. It provides a simultaneous reading, which can allow the user to help identify the potential family that the suspect gas belongs to.

Corona Discharge Ionization

Several attempts have been made to use corona discharge (CD) as the ionization source in emergency responder detection devices and now are common in terrorism agent detection. Corona discharge is also known as plasma chromatography. It is the most likely candidate to replace the nickel 63 (^{63}Ni) radiation source found in common IMS devices; it is the detection source in the LCD device manufactured by Smiths Detection.

A CD sensor has two electrical electrodes that, when energized, create an electrical arc between the electrodes. The arc, composed of electrical energy, has the ability to ionize the gas sample. One of the advantages of CD is that, by changing the amount of current applied to the electrodes, the sensor can change the ionization potential. The CD sensor has better detection limits and range than its radioactive cousin when it develops ions that are positive. When negative ions are formed, the CD sensor performs worse than the ^{63}Ni source. Manufacturers have had to overcome this obstacle. Other advantages of a CD sensor a detection range of 0.1–10,000 ppm and in some cases even lower. The CD sensor can ionize materials, much like a FID, and can cover chemicals more than an 11.5 eV ionization potential.

Gas Chromatography

Gas chromatography (GC) is a possible option for response teams to take after the use of handheld devices. Several companies produce field instruments that have some application in emergency response (**FIGURES 7-6** and **7-7**). The devices used for emergency response and in the laboratory combine the gas chromatograph and the mass spectrometer (GC/MS). One device uses a new mass spectrometry technology and does not use the GC portion of the analysis.

The GC uses a heated column through which a sample is passed in the presence of an inert gas such as helium or nitrogen. Each component of the sample has an attraction for the material that makes up the column and has a specific holdup and travel time within the column. A graphical picture of travel times and relative concentration is compared in the library, and a match of components can hopefully be made. With multiple gases

IONIZATION DETECTION DEVICE COMPARISON

Chemical	Ionization Potential (eV)	PID 10.6 eV Lamp	PID 11.7 eV Lamp	FID	Corona Discharge
Acetone	9.69	X	X	X	X
Acetylene	11.4		X	X	X
Ammonia	10.18	X	X		X
Ethyl ether	9.53	X	X	X	X
Formaldehyde	10.88		X		X
Freon 112	11.30		X	X	X
Hydrazine	8.93	X	X	X	X
Hydrogen cyanide	13.60			X	X
Hydrogen fluoride	15				X
Hydrogen peroxide	10.54	X	X	X	X
Hydrogen sulfide	10.46	X	X		X
Methane	13.0			X	X
Methyl alcohol	10.84		X	X	X
Methyl ethyl ketone	9.54	X	X	X	X
Nitrous oxide	12.89			X	X
Phenol	8.5	X	X	X	X
Phosgene	11.55		X	X	X
Propane	11.07		X	X	X
Sulfur dioxide	12.30			X	X
Toluene	8.82	X	X	X	X
Triethylamine	7.5	X	X	X	X
Xylene	8.56	X	X	X	X

present, each will "boil off" at differing temperatures, causing an individual peak or spike on a spectrum (much like the heartbeats on a heart monitor). Each chemical has a retention time and produces a peak at a specific time interval, which identifies the component. The length and width of this peak can be used to determine the relative amount of each component material when the spectrum is compared against a known standard (**FIGURE 7-8**). The sampling time varies depending on the volatility of the whole sample but can often be

OTHER TECHNOLOGY

In addition to corona discharge, another new ionization method is atmospheric pressure ionization (API), which is most commonly associated with liquid chromatography/mass spectrometry. API involves the development of ions at atmospheric pressure by means of two methods: (1) Atmospheric pressure chemical ionization (APCI) is used for gas phase materials, and (2) electrospray ionization (ESI) is used for liquids. The APCI method involves heating the sample material and introducing a carrier gas, such as nitrogen. The mixture then moves to a corona discharge, which creates new ions. The ESI method, which may also be called desorption electrospray ionization (DESI), takes the sample material in solution and under pressure sprays it into the corona discharge, resulting in ions being formed. Many such methods are being utilized to allow for quicker GC/MS testing, which can take a considerable amount of time. These methods reduce the testing time to minutes from hours and, given enough interest, may become stand-alone detection methods.

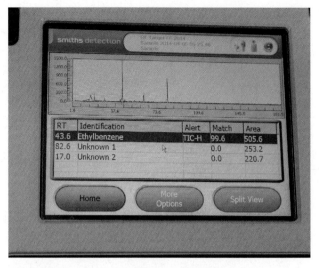

FIGURE 7-8 Spectra from a GC/MS.
Courtesy of Christopher Hawley.

FIGURE 7-6 A Inficon Hapsite GC/MS.
Courtesy of Christopher Hawley.

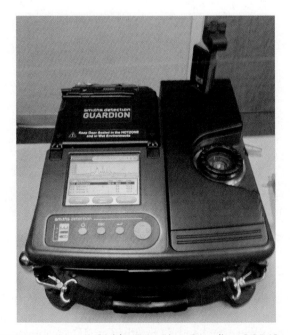

FIGURE 7-7 A Smiths Detection Guardion GC/MS.
Courtesy of Christopher Hawley.

completed in 3–20 minutes. GC is one of the last resorts when dealing with the otherwise unidentified, but the lab must have a comparison sample of the gas or liquid you want them to analyze before they can even begin to hazard a guess. A GC cannot detect water, so if the sample contains water, the other components will be overstated unless the water content is independently measured and factored into the analysis calculations. Field GCs are available, but they are expensive and do require regular upkeep to keep them operating. Although this technology has changed a great deal in the past few years, it has become much easier to use than it had been. There still is a need for a considerable amount of training to master the use of this device, and the unit needs to be run and calibrated at least weekly.

Mass Spectrometry

The mass spectrometer (MS) is usually coupled with a GC, and it is really a seamless operation with the two technologies (**FIGURE 7-9**). The MS of a GC/MS is the actual identifying portion of the device. The GC separates the molecules, as discussed previously, and then the MS portion provides the identification. To accomplish this identification, the MS measures the relative mass of the molecular fragments and compares them to the mass of a comprehensive list of materials in its library, as each molecular fragment has a different weight. To identify a material, the footprints must be in the library. This library is the most important part of the GC/MS, and the more extensive the library, the more successful you will be at identifying an unknown substance. The 908 Devices Mass Spectrometer MX908 uses high-pressure mass spectrometry (HPMS), which is a relatively new MS technology and is not coupled with

FIGURE 7-9 A MX908™ handheld high-pressure mass spectrometer produced by 908 Devices inc.
Provided by 908 Devices.

FIGURE 7-10 Devices should be able to be used when in protective clothing. This MX908 high-pressure mass spectrometer is designed for this hot zone use.
Provided by 908 Devices.

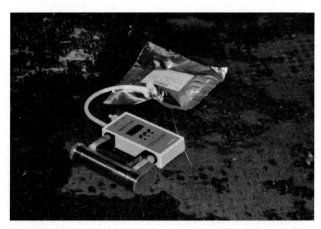

FIGURE 7-11 A Tedlar bag attached to a gas pump.
Courtesy of Christopher Hawley.

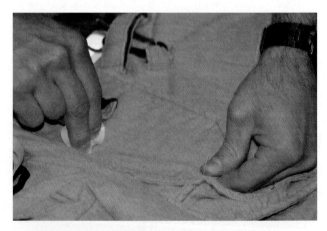

FIGURE 7-12 Swabbing a clean set of turnout gear to check the efficacy of the cleaning process.
Courtesy of Christopher Hawley.

a GC (**FIGURE 7-10**). The name is a slight misnomer as it implies that there is a high-pressure analysis going on in the device, when in reality the operating pressure in the device is just higher than other MS devices and just a bit lower than atmospheric pressure. The device can analyze materials in vapor mode or accept wipe samples to look for trace amounts of liquids or solids. GC/MS and MS devices operate in a vacuum situation, but the others require large vacuum pumps to operate. The distinct advantage of HPMS is size; the detection system has been miniaturized and is handheld. GC/MS devices have the two detection systems and carrier gas in addition to the electronics that manage the software and a battery. The MX908 device is only the MS portion and does not require any carrier gas, and the MS detection sensor is small. It can identify many materials, if the material is in the instrument's library. The MX908 library focuses on materials of concern, such as chemical warfare agents, explosives, illicit drugs, and their precursors. The MX908 software automatically searches the onboard library for possible matches and does not require the user to read a spectrum.

With some of the GC/MS devices there is some setup time required, and the sampling tools need to be prepared. The device must also warm up and ready itself. The Smiths Guardion device warms up in 5 minutes, while some devices take at least 20 minutes to warm up. The mechanism to run a sample will vary with the state of matter of the material being sampled. Some instruments, such as the MX908, can run in continuous vapor sampling mode (similar to a PID) and do not require the user to run a specific sample. **FIGURE 7-11** shows a Tedlar bag attached to a Smiths Gas-ID sample pump is shown collecting an air sample. In **FIGURE 7-12**,

FIGURE 7-13 A swab in a sample jar with a solvent being prepared for analysis.
Courtesy of Christopher Hawley.

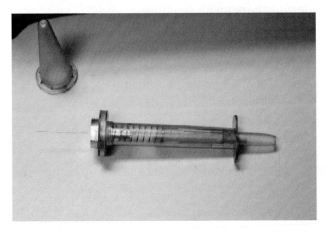

FIGURE 7-14 A solid phase microextraction (SPME) fiber syringe used to grab sample for insertion into the GC/MS.
Courtesy of Christopher Hawley.

a swab is being used to collect samples, which will be placed in a clean sample jar that has an organic solvent in it (**FIGURE 7-13**). The third option is to use the solid-phase microextraction (SPME) fiber syringe directly, which is then placed into the GC/MS (**FIGURE 7-14**). When introducing a sample into a GC/MS or MS, it's important not to overload the instrument. The units can easily become saturated and may require numerous runs to clear out the material. Many GS/MS and MS instruments have a protocol for diluting samples prior to introducing them into an instrument. When desiring to use the GC/MS, it requires some setup and thought, especially the need to develop a sampling plan. A potential tactic to include in a sampling plan is to use a PID or FID to determine the presence or location of an unknown material, then use a GC/MS or MS to identify the material. Tracking of the samples requires some thought as well, as several samples can overwhelm the operator and make it a challenge for him or her to keep track of the samples.

The one consideration is that the GC/MS will identify the various components but will not identify what the product is. As an example, if you run a sample of gasoline, the GC/MS will report back that it has located the various components of gasoline, which can be hundreds or more. Operating and interpretation of GC/MS can be challenging. Teams that deploy a GC/MS should consider having a proficiency program as part of the recurring training to ensure accurate results. If your team cannot afford a GC/MS, then it is recommended that you locate a local lab that will run samples for you. Some local police department crime labs have GC/MS capability, although they are often set up to do drug screens. Other industrial locations usually have labs for quality assurance, and you should make arrangements to have samples run in an emergency. In most cases, county or state environmental departments have labs that can run samples, but prior arrangements also need to be completed. When sending samples out, you need to determine what safety screening is required and what the lab prefers in the way of type of sample, sampling medium, and sample size. As litigation is always possible, it is best to split your samples between two labs and compare the results, and always be able to document the chain of custody.

After-Action REVIEW

IN SUMMARY

The use of ionizing detection devices is essential for responder safety, as they are valuable at measuring the potentially toxic portion of risk-based response. Most chemicals that hazardous materials teams respond to are flammable or combustible. Some of these chemicals are toxic (a few severely so) long before they reach their fire risk. The sensitive nature of the newer parts-per-billion models will provide valuable clues to possible chemical hazards during responses. Learning what types of materials can be measured, what levels can be expected, and the limitations of the detection devices are all crucial in the successful use of ionizing detectors.

KEY TERMS

Corona discharge (CD) A form of ionization that uses an electrical arc to ionize a gas sample.

Flame ionization detector (FID) A device that uses a hydrogen flame to ionize a gas sample; used for the detection of organic materials.

Gas chromatograph/mass spectrometer A detection device that separates mixtures and identifies them by their retention and travel times, then ionizes the separated chemicals and compares the ions to a library of known materials.

Gas chromatograph (GC) A detection device that separates mixtures and tentatively identifies them by their retention and travel times.

Ionization potential (IP) The ability of a chemical to be ionized or have its electron removed. To be read by a PID, a chemical must have a lower IP than the lamp in the PID.

Mass spectrometer Almost always coupled with a GC, the MS ionizes chemicals and measures the weight of the given ions and compares it to a library of known materials.

Photoionization detector (PID) A detector that measures organic materials and some inorganics in air by ionizing the gas with an ultraviolet lamp.

Review Questions

1. What is ionization?

2. What does the PID use as its ionization source?

3. What must be known about a chemical to determine if a PID will detect it?

4. What is a disadvantage of a FID device?

5. What is ionization potential?

6. What is the maximum ionization potential material that can be read by a FID?

7. What is the most typical PID lamp, and why is knowing its strength so valuable?

CHAPTER 8

Colorimetric Tube Sampling

OUTLINE

- Learning Objectives
- Introduction
- Colorimetric Tube Science
- Detection of Known Materials
- Detection of Unidentified Materials
- Street-Smart Tips for Colorimetric Tubes
- Additional Colorimetric Sampling Tips
- In Summary
- Key Terms
- Review Questions

LEARNING OBJECTIVES

Upon completion of this chapter, you should be able to:

- Describe colorimetric science
- Describe common colorimetric detection devices
- Describe the process for the detection of known materials
- Describe the process for the detection of unidentified materials

Hazardous Materials Alarm

Ammonia and Tube Colors—Most of my unusual incidents with colorimetric tubes have involved ammonia. In one incident, we responded to a reported ammonia leak at a chicken processing plant. I was setting up the tubes in preparation to go into the building when I snapped the ends of the ammonia tube, and it instantly changed colors. We quickly changed the isolation distance another block or two. In another case, we had gone to assist an engine company with a cylinder buried under an apartment complex. The woman who lived in the basement apartment had been smelling ammonia, and the rental company had been trying to locate the source. They had torn the woman's apartment apart, including the walls, floor, and ceiling. When they did not find anything in the apartment, they decided to dig near the foundation, where they found the cylinder buried under the foundation. The apartment building had been there at least 21 years, so the cylinder had been there at least that long. The cylinder looked like a nurse ammonia tank a farmer would have used. The apartments were built on an old farm, so it is suspected that the cylinder was left behind from the farm and was buried with the other backfill. We grabbed a colorimetric tube for ammonia and proceeded to the hole, with my partner jumping in. He started pumping the colorimetric tube, and very quickly I asked him, when did it change? His reply was that it had not changed. I was not sure what color it was to begin with, so I grabbed a new tube to compare. It had changed colors, but we did not know when, so I prepared another tube and handed it to my partner, and on the first pump stroke, it went off scale. With all the information, we were pretty sure the cylinder was ammonia. We dug the cylinder out and used the bomb technicians' trailer to move the cylinder to a remote location where we opened the cylinder in a porta-tank of 2500 gallons (9,464 liters) of water and made a large amount of ammonium hydroxide solution.

1. Regarding detection devices, where would you start with trying to determine what was in the cylinder?

2. What would you use to determine if your readings were safe or not?

3. How large of an isolation area would you have?

Introduction

Colorimetric tubes are used for detecting known and unidentified gases or vapors and come in a standard kit (**FIGURES 8-1** and **8-2**). Colorimetric tubes are an easy method to narrow down an unidentified material to its chemical family. With an experienced user, specific identification can be made. It is unfortunate that this valuable tool in risk assessment is often overlooked or used incorrectly. Colorimetric tube sampling is not a simple task, and the user must be familiar with every aspect of the unit being used. The results one can obtain are invaluable. Even if the responder cannot determine what material is present, colorimetric tubes can tell us what is not present, which at the time can be as important.

FIGURE 8-1 Sensidyne kit.
Courtesy of Sensidyne.

FIGURE 8-2 Colorimetric tube set.
Courtesy of Christopher Hawley.

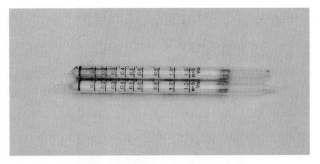

FIGURE 8-3 A bellows style pump.
Courtesy of Christopher Hawley.

LISTEN UP!

Colorimetric tubes are an easy method to narrow down an un-identified material to its chemical family.

Colorimetric Tube Science

Colorimetric tube sampling consists of placing a glass tube filled with a **reagent** into a pumping mechanism, which causes air to pass through the reagent, a powder, or crystals at a calibrated rate (**FIGURE 8-3**). The pumping mechanism can include bellows pumps, piston pumps, and thumb pumps. The type of pump used determines the pump stroke volume, stroke interval, and size of the tube. These pumps are used to draw 50–100 mL of air across the tube (**FIGURE 8-4**). If the gas or vapor reacts with the reagent, a color change should occur,

indicating a response to the gas sample. The three areas of change that are found are change of color, length of change, and color intensity. Detection tubes are made for a wide variety of materials and generally follow the chemical family lines (i.e., hydrocarbons, halogenated hydrocarbons, acid gases, amines). Although the tubes may be marked for a specific gas, they usually have cross-sensitivities (react to other materials), which at times is the most valuable aspect of colorimetric sampling. The tube systems may also have extension hoses, pyrolizers (heaters), prefillers, conditioning layers, sealed internal ampoules, pretubes, and reagent tubes. It is very important that responders reading the instructions perform the sampling correctly with all the appropriate parts. Because colorimetric tubes rely on chemical reactions, the additional components to the tube aid in the detection of many materials. The pyrolizer creates a high-temperature flame, thereby creating a high-temperature environment inside the tube, which speeds up the chemical reaction. The sealed inner ampoules are for materials used to enhance the chemical reaction, once the sample gas is pulled through the tube.

One colorimetric system that has been developed by Dräger involves the use of bar-coded sampling chips. This system, called the **chip measurement system** (CMS), involves the insertion of a sampling chip into a pump (**FIGURE 8-5**). The pump recognizes the chip in use and provides the correct amount of sample air through the reagent. By using light transfer through a reagent-filled ampoule, the pump provides an accurate reading of the gas that may be present. The advantage to this system over regular colorimetrics is that the reading

FIGURE 8-4 A piston style pump.
Courtesy of Sensidyne.

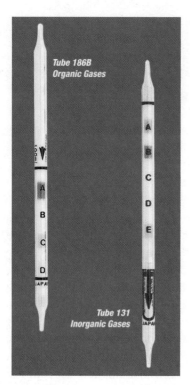

FIGURE 8-5 Sensidyne 131 and 186B tubes.
Courtesy of Sensidyne.

is provided on a liquid crystal display (LCD) screen in parts per million, so it does not need interpretation. The other colorimetric systems require a length and color interpretation by a human, something that is at times faulty. The disadvantage to the CMS is that a general sensing tube is not available yet, nor can it sample more than one material at a time. Some standard colorimetric systems have the capability to increase the sensitivity of a tube or to reuse a tube after negative results, but the chip system does not allow this. The chips have 10 sampling capillaries, which after each is opened cannot be reused, nor can the sampling range be enhanced.

> **LISTEN UP!**
>
> Even if the responder cannot determine what material is present, colorimetric tubes can tell us which ones are not present, which at the time can be as important.

Detection of Known Materials

If you know the material for which you are sampling, there is a good possibility that a tube is available for that material. Most tubes are set up to read in parts per million, while others may be in percent by volume, and, even less common, some tubes just indicate presence. Most manufacturers provide a variety of sampling ranges

for responders' use and a variety of sampling devices is available. When only one set of tubes is available, it is possible to further enhance this range by increasing or decreasing the amount of sample air that passes through the tube. With all colorimetric sampling, make sure to read the instructions, and keep in mind that these tubes are calibrated for a certain temperature and humidity. The manufacturer provides temperature and humidity correction factors that should be used when using the tubes outside their normal operating parameters.

> **LISTEN UP!**
>
> When looking for an unidentified material, knowledge about colorimetric sampling, chemistry, and street sense is essential.

> **SAFETY TIP**
>
> Using colorimetrics, we can come close to identifying the chemical family, its characteristics, and the potential hazards of that material. Even more important, we can determine whether it is immediately dangerous to life and health (IDLH).

Detection of Unidentified Materials

Detecting unidentified materials is not an easy task. When looking for an unidentified gas or vapor, knowledge about colorimetric sampling, chemistry, and street sense is essential. It takes more than one person to complete an investigation of an unidentified material, one to sample, one to hold tubes, one to track results, and one to figure sampling strategy. Most manufacturers provide a flow chart that provides a guideline to follow when looking for unidentified materials. The flow charts guide responders to a chemical family and hopefully to narrow the choices of tubes and possible chemicals present. There is no one monitor that exists that will absolutely 100 percent indicate that "x, y, z" is the material present. However, by using colorimetrics, we can come close. We can identify the chemical family, its characteristics, and the potential hazards of that material. Even more important, we can determine whether it is immediately dangerous to life and health. One of the best things that a colorimetric system does is rule out what is not present. If you utilize an entire standard colorimetric tube set, you have sampled for a wide variety of chemical hazards and have effectively covered the whole spectrum. **TABLE 8-1** provides an overview of an unidentified

TABLE 8-1 Unidentified Flow Chart for Colorimetric Tubes

Dräger Unidentified Flow Chart

Tube	Chemical Family	Reaction/Notes
Polytest	Organics and inorganics	Positive means that something is present; the higher the reading, the more material that is present. A negative reading does not mean the air is safe.
Perchloroethylene	Halogenated	Most common would be Freon, but others include methyl chloride, dichloromethane, trichloromethane, chloroform, and carbon tetrachloride.
Phosgene	Phosgene	Phosgene
Ethyl acetate	Organic substances	Rules out inorganics
Benzene	Aromatic hydrocarbons	Benzene, toluene, ethyl benzene, xylene, cumene, and styrene
Acetone	Ketones	Acetone, methyl ethyl ketone, methyl isopropyl ketone, and methyl isobutyl ketone
Alcohol	Alcohols	Includes isopropyl alcohol, ethyl alcohol, methyl alcohol, ethylene glycol, propylene glycol, propyl alcohol, 1-propanol, 2-propanol, allyl alcohol, 1-butanol, 2 butanol, tert-butyl alcohol
Ammonia	Amines and ammonia	Amines, ammonia, and hydrazine
Formic acid	Organic acid gases	Color change indicates type of acid that may be present.

Sensidyne/Kitagawa Colorimetric Tubes

Tube	Chemical Family	Reaction/Notes
186B	Organics – acetaldehyde, acetone, acetylene, aniline, benzene, 1,3-butadienem butane, 1-butanol, butyl acetate, carbon disulfide, cresol, ethyl acetate, ethyl amine, ethyl benzene, Cellosolve, ethylene, ethylene oxide, formaldehyde, gasoline, heptane, hexane, isopropyl alcohol, kerosene, methyl alcohol, methyl ethyl ketone, methyl isobutyl ketone, methyl mercaptan, pentane, phenol, propane, styrene, tetrachloroethylene, tetrahydrofuran, toluene, 1,1,1-trichloroethane, trichloroethylene, vinyl chloride, and xylene. Also the following inorganics: arsine, carbon monoxide, and hydrogen sulfide	See the Sensidyne charts in Figure 8-6A for more information. The charts that are included with the kit also provide some guidance as to the quantity of material that is present.

(continued)

TABLE 8-1 Unidentified Flow Chart for Colorimetric Tubes (*continued*)

Tube	Chemical Family	Reaction/Notes
131	Inorganic gases—acetic acid, amines, ammonia, carbon monoxide, chlorine, hydrogen chloride, hydrogen sulfide, nitrogen dioxide, phosphine, and sulfur dioxide. Also the following organic gases: acetylene and methyl mercaptan	See the Sensidyne charts in Figure 8-6B for more information. The charts that are included with the kit also provide some guidance as to the quantity of material that is present.

Gastec Colorimetric Tubes		
Tube	**Chemical Family**	**Reaction/Notes**
Polytech I	Carbon disulfide Hydrogen sulfide Carbon monoxide Acetone Acetylene Ethylene Benzene Propane, propylene Styrene Trichloroethylene Gasoline Toluene, xylene	No differentiation is possible with some of these combinations, although among the major groups, there are color differences. As an example, both carbon disulfide and hydrogen sulfide cause a green color change.
Polytech II	Ammonia, amines Sulfur dioxide Hydrogen sulfide Carbon monoxide	Also does hydrogen chloride, chlorine, nitrogen dioxide, and phosphine
Polytech III	Ammonia Hydrogen sulfide Hydrocarbons	Also does sulfur dioxide, hydrogen chloride, chlorine, nitrogen dioxide, butane, gasoline, and liquid petroleum gas (LPG)
Polytech IV	Ammonia, amines Hydrogen chloride Hydrogen sulfide Chlorine Carbon monoxide Carbon dioxide	Also does sulfur dioxide, nitrogen dioxide, acetylene, ethylene, phosphine, hydrogen, methyl mercaptan, and propylene

Dräger Polytest Note

The Polytest tube relies on an iodine reaction with a suspected contaminant. In Chapter 10, we discuss the use of the iodine chip test from the HazCat kit. For street purposes, the Polytest tube is chip test in a tube. The color reaction in the tube can provide some initial guidance to the chemical family. Chapter 10 has more information on these colors, but an example would be yellow or light orange indicates polar hydrocarbons such as acetates, alcohols, ethers, esters, nitriles, acids, or a mixture.

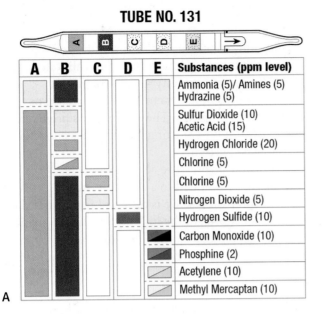

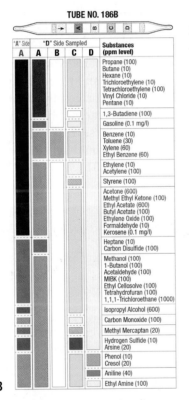

FIGURE 8-6 (A, B) Sensidyne color charts.
Courtesy of Sensidyne.

FIGURE 8-7 Dräger Multitube system which can test five different chemicals or groups of chemicals at one time.
Courtesy of Christopher Hawley.

flow chart from three manufacturers. The Dräger system requires several tubes to work through the various families. The Sensidyne and Kitagawa system uses two tubes that can identify more than 70 possible gases within a few minutes (**FIGURES 8-6** and **8-7**). The Gastec system has four **Polytech** tubes that are used to provide "hits" for possible substances. The Polytech I tube is to be used much like the Dräger **Polytest** tube, following their unidentified flow chart, which somewhat resembles the Dräger chart but has some variations and different

tube names. A street-smart tip: Use the Sensidyne or Kitagawa tubes to start or the Gastec Polytech I and Polytech IV to start your search, then follow the search pattern provided in Chapter 11 for unidentified materials. Both Dräger and Mine Safety Appliances Company (MSA) have multitube testing systems that can test up to five materials at a time.

The one major key point when using colorimetric tubes is to be familiar with the instructions that come with each tube. No one can memorize the instructions for each tube and its varying requirements and problems. Responders should know how to obtain this information from the instructions with each tube. It is recommended that the department make a book of instruction sheets for the tubes they carry to make searching easier. The instruction sheet lists the number of pumps or strokes for the tube, the color change to be anticipated, and, most importantly, the cross-sensitivities and associated problems with the tube. When sampling for unidentified materials, cross-sensitivity is most important, as it may be your only clue as to the material present. The tubes are manufactured with filters, screens, and other materials at each end. A material for which the tube is sensitive may only change the filter and not change any other part if the tube. If the responder misses this change, it may be the only "hit" for that material. If the responder gets a change on a tube that is not indicated by the instructions (either in the indicating layer information or the cross-sensitivities listing), one can only assume that there is something present, but identification by this tube is not possible. Only rely on color changes that are indicated somewhere on the instruction sheet.

Street-Smart Tips for Colorimetric Tubes

The instructions with colorimetric tubes and some other texts mention that the tubes can be inaccurate up to ±25 percent of the indicated reading, sometimes even higher. This inaccuracy rate varies tube to tube, and even the instructions provide a varied rate. Most tubes have a shelf life of 1–3 years. Budgeting should be planned to replace expired tubes. During testing, some tubes that are years out of date and have been stored in unusual conditions have been found to be up to 98 percent accurate, but for liability reasons, it is best to follow the manufacturer's provided variance factor. RAE systems by Honeywell sponsored a test of colorimetric tubes from various manufacturers. The test was completed by three independent laboratories. This testing found that the tubes were accurate within 15 percent, less than the manufacturers listing of 25 percent. **TABLE 8-2** shows an excerpt from some of the testing results. Many factors can cause this inaccuracy, some of which are provided in the following list:

- Humidity plays a major factor in the response of some tubes, and most tubes are calibrated to 50 percent humidity. If the humidity is different, you must factor in the increased or decreased humidity. The instructions provide the calculation for adjusting for humidity above or below 50 percent.
- Atmospheric pressure also plays a role, as the tubes are calibrated to a certain pressure. If the atmospheric pressure is different from what the tube is calibrated, the instruction sheet provides a formula to determine the actual reading. Dräger provides a correction factor formula for atmospheric pressure. You would multiply the correction factor by the reading on the tube. To obtain the correction factor, you divide standard atmospheric pressure by the actual atmospheric pressure. Standard atmospheric pressure is 14.7 psi, which is equal to 1013 hPa, 1 ATM, 30 inches Hg, or 1 bar. If you get a reading of 100 ppm on a tube and the atmospheric pressure is 30.71 inHg, the correction factor would be 0.977. The actual reading should be 97.69. If the atmospheric pressure was 29.6 inHg, the correction factor would be 1.01, and the actual reading would be 101.

TABLE 8-2 Accuracy of Colorimetric Tubes

Tube	Challenge Chemical	Resulting Reading
RAE 103-10; 2–50 ppm	25 ppm H_2S	29 ppm
Gastec 4LL; 2.5–60 ppm	25 ppm H_2S	27 ppm
Kitagawa 120SD; 1–60 ppm	25 ppm H_2S	29 ppm
Dräger H_2S 2/b; 2–60 ppm	25 ppm H_2S	30.5 ppm
MSA H_2S-1; 1–200 ppm	25 ppm H_2S	26 ppm
RAE 104-50; 1–20%	10% CO_2	9.2%
Gastec 2H; 1–10%	10% CO_2	9.6%
Kitagawa 126SH; 1–20%	10% CO_2	9.8%
RAE 103-10; 2.5–60 ppm	10, 20, and 40 H_2S	10, 20, and 46 ppm
RAE 104-40; 0.05–1%	0.20, 0.50, 0.80% CO_2	0.17, 0.49, and 0.74%

- Each tube has a specific operating temperature range; most are good for 32–122°F (0–50°C).
- Age, light, and storage conditions also determine the accuracy of the tubes. All tubes have a specific shelf life, and because the tubes work on a chemical reaction and corresponding color change principle, light can cause a change within the tube. Although some manufacturers recommend that the tubes be stored in the refrigerator, if you will be taking them in and out frequently, it is best to store them in conditions that do not have temperature extremes.
- One of the most common mistakes when using tubes is not waiting for the pump stroke to finish. All pumps require 30 seconds to a minute per pump stroke on average. Some may take longer and may take up to 5 minutes per air draw. The instructions provide the length of time for the pump stroke. All pumps have an end-of-flow indicator, which alerts the user when the air draw is complete. Before drawing more sample into the tube, you must wait until the previous draw is finished.
- Colorimetric tubes are gas and vapor detection devices. The higher the vapor pressure, the easier it will be to characterize the unknown. Dräger sells a sampling tube for sulfuric acid. The vapor pressure for sulfuric acid is significantly low. The tube requires 100 pump strokes, and they each take a minute. The sampling time will be more than 100 minutes. In addition, at the end of the sampling process, there is an inner ampoule that must be broken, and you use a one-quarter pump stroke to draw the ampoule liquid into the detection area. The resulting color change must be read immediately as it will disappear. Sensidyne has a sulfuric acid tube as well, and it requires five pulls of the pump piston and takes 100 seconds. Even with this quicker method, there are other detection methods. As sulfuric acid is a liquid, pH paper would indicate that it is an acid with a very low pH and no vapor pressure.
- More than one gas present in a sample area can alter the readings. Remember that the tubes react to many other gases than just what the box indicates. It is important to know the cross sensitivities of each tube you use.
- It is important to know what the color of the tube is prior to using it in the pump so you can recognize a color change. Having the sampling team keep an unused, closed tube for comparison is a good idea.
- Visible vapor that is released from various tubes is usually water vapor and is harmless, but in some cases, other gases may be released. It is not uncommon for tubes to heat up and release visible vapor, depending on the gas being sampled.
- There are chemical tubes that create an extremely hot chemical reaction, which in a flammable atmosphere is a perfect ignition source. When using colorimetric tubes, it is important to check for the presence of flammable gases prior to using some colorimetric tubes. Tubes that could cause problems are rare, but two common tubes that heat are the Dräger halogenated hydrocarbons tube and the Gastec oxygen tube. With some off gases, persons with respiratory problems could be affected.
- It is important to clear the pump between tubes to clear the previous sample. As chemical reactions take place within the tubes, it is essential to ensure a clean pump for the next sample.

Additional Colorimetric Sampling Tips

By increasing or decreasing the amount of air passing through the tube, you can increase or decrease its sensitivity, and some examples are provided in the Hazardous Materials Alarms Ammonia and Tube Colors. By changing this sensitivity, you do not need to maintain several boxes of tubes. Even when you do not get any change on a tube, by changing its sensitivity, you may get a "hit" for that gas. As an example, the ammonia tube takes 10 pumps and measures 2–30 ppm. If after pumping 10 times you did not get any reading, the number of pumps or strokes can be increased by another 10, and that will increase the sensitivity to a range of 1–15 ppm. When sampling in this fashion, any readings are estimated but are valuable when searching for an unidentified material. Many times, small amounts of chemicals in the air present a chronic hazard and are irritating over an extended period. However, if pumping the ammonia tube 10 times results in a reading that is off the scale, then we can decrease its sensitivity by limiting the number of pump strokes. If we pump the tube five times, and the tube is reading 4–60 ppm, pumping the tube one time would allow a reading of 20–300 ppm. In some cases, the color change is very quick, so it is important to know what the original color of the tube was. For some chemicals, the tube will change dramatically without any air being pumped through the tube and will provide a quick response to those gases. The instructions must be read to ensure that there are no special restrictions on the tube used in this fashion. Some of the chemical warfare agent tubes have inner ampoules that are broken after the air is passed through the tube and could not be used in this fashion.

It is important to know some basic chemistry, as this is invaluable when dealing with unidentified materials in the identification flow charts. In the event you did not get any hits on the flow chart, your next option is to go through every tube in the kit, watching for any possible hits. In Chapter 11, there are some example sampling strategies using colorimetric tubes. It may be best to combine tubes from the various manufacturers in your system. By combining several tubes, you can shorten the time required to do some broad-range sampling. Although there has been considerable discussion in the response community about using one manufacturer's tube in another manufacturer's pump, the best rule is that a fat tube goes with a pump intended for fat tubes, and a skinny tube is used in a pump for skinny tubes. Most tubes require 100 cc of air to be passed across the sampling media, and the tube does not care if that is pulled across through a piston pump or a bellows pump. When trying to identify an unidentified gas in a characterization mode, the pump type is a moot issue. When doing occupational exposure, evidentiary, or other legal sampling, then it is advisable to use the manufacturer's recommended pump. Because the sensitivity varies with colorimetrics, a slight change is a possible clue. Also, a color change in a prefilter is a hit and may indicate a possible gas. When looking for an unidentified gas, many times we can find unusual sources, so check for all possible gases available. When you have gone through the whole sampling process and have found nothing, what you can say is: This, this, and this were not present at the time of sampling.

After-Action REVIEW

IN SUMMARY

Although colorimetrics are one of the more confusing detection devices, they can provide the greatest amount of information. They can take an unidentified situation and bring it to a chemical family, if not to a specific identification. They are time consuming and require the user to read the instructions. No one colorimetric system is better than the other, and, in fact, the best system is a combination of several manufacturers' systems.

KEY TERMS

Chip measurement system (CMS) A colorimetric sampling system that uses a bar-coded chip to measure known gases.

Colorimetric A form of detection that involves a color change when the detection device is exposed to a sample chemical.

Colorimetric tubes Glass tubes filled with a material that changes color when the intended gas passes through the material.

Polytech The name for Gastec's unknown gas detector tube.

Polytest The name for the Dräger unknown gas detector tube.

Reagent A chemical material (solid or liquid) that is changed (reacts) when exposed to a chemical substance.

Review Questions

1. What are colorimetric tubes used to identify?

2. What is an important item to review prior to the use of colorimetric tubes?

3. After you test with colorimetric tubes and find nothing, what is known?

4. When you are sampling for unidentified materials, what aspect of the tubes can assist in a possible identification?

5. The unknown flow charts are broken down into which groups?

6. How can you make a colorimetric tube more sensitive?

7. In between tubes, what is an important task?

CHAPTER 9

Radiation Detection

OUTLINE

LEARNING OBJECTIVES

Upon completion of this chapter, you should be able to:
- Identify sources of radiation
- Describe the basis of radiation
- Identify types of radiation
- Describe radiation dose and its impact on humans
- Identify the common types of radiation detection devices

Hazardous Materials Alarm

While doing some radiation detection training, we hid four radiation check sources in a room. We sent the first entry crew into the room to find the sources, and when they were finished, they were to report back to us. The crew came back out a few minutes later and said we found all five sources. We looked at each other, and I asked the other instructor to confirm that we had only hidden four sources. She confirmed that indeed we had only hidden four. We asked the crew to show us where the five were located. They showed us the four we had hidden, and then they showed us a location on a wall where they were getting readings. This was a concrete wall. It was pretty thick, and it was indeed emitting radiation. It was not at harmful levels but did warrant some investigation. Luckily, the students were from the emergency situations unit, and we told them they needed to investigate the source. The investigation showed that some irradiated steel had been used in the construction, and it was later removed.

1. What method would you use to determine if radiation were present?

2. How do you determine if there are harmful levels of radiation present?

3. What is the normal level of background radiation that you would anticipate finding in your response area?

Introduction

Dealing with radiation incidents makes many responders uncomfortable. Emergency responders do not have an in-depth understanding of radiation and its applicable hazards. With continued concern of terrorism, responders need to be more proficient with the detection and monitoring for radiation. Although **nuclear detonation** is very unlikely, the use of a **radiological dispersion device (RDD)**, in which a conventional explosive is used to distribute radioactive materials, is not beyond imagination. Any number of potential sources of radiation could easily be used for this purpose. Therefore, responders should be more skilled in handling radiation events so that the community does not panic.

FIGURE 9-1 A radiation detection device.
© Jones & Bartlett Learning. Photographed by Glenn E. Ellman.

Radiation detectors for emergency responders are divided into two major groups: One measures exposure to radiation, and the other measures the current amount of radiation in the area (**FIGURES 9-1** and **9-2**). To effectively measure radiation, at least two detectors are needed as no detector measures all five types of radiation.

FIGURE 9-2 A radiation pager/dosimeter.
© Jones & Bartlett Learning. Photographed by Glenn E. Ellman.

Sources of Radiation

We are subjected to radiation exposure in various forms every day. Our bodies contain radioactive substances. We eat foods containing radiation, and we breathe in radiation without any harm every day. In fact, our exposure to everyday radiation sources far exceeds what we would experience if we worked at a nuclear power plant. Through television, medical tests, and elevation, we are subjected to levels of radiation that cause us no harm under normal circumstances. To understand how radiation can harm us, we must understand radioactivity, not only in terms of how radiation affects the human body but also how radiation monitors work.

To explain radioactivity, we must describe the basic atom, specifically the nucleus. An atom is comprised of electrons, neutrons, and protons. Protons and neutrons reside in the nucleus at the center of the atom, while negatively charged electrons orbit the nucleus. Protons have a positive charge and determine the element or type of atom. Neutrons are the same size as protons but are neutral. Each element, with a given number of protons, can assume several forms or isotopes, which are determined by the number of neutrons in the nucleus. The chemical properties of each isotope of an element are the same; except for separating the isotopes in a centrifuge, you cannot chemically or physically identify one from the other. If there are too few or too many neutrons, however, the nucleus becomes unstable. Radioisotopes, isotopes whose nuclei are unstable, are radioactive and emit radiation to regain stability. This emission of radiation, known as **radioactive decay**, usually takes the form of **gamma (γ)** radiation but may also be **alpha (α)**, **beta (β)**, or **neutron** radiation, as shown in **TABLES 9-1** and **9-2**. An unstable

In a small community, near the end of the school year, a high school demonstration of a radiation monitor resulted in an interesting response to a radioactive rock. The teacher was checking the radiation levels of various items, and the monitor started to click loudly when placed near a large rock that had been in the classroom for many years. The hazardous materials team was called and, wearing lead aprons, placed the suspect rock into a lead pig. The school was closed for the remainder of that day and the next, and the children were advised to take showers and wash their clothes and shoes. The rock, which was from an unknown location, contained uranium, a common material. The issue at hand is that the detector used was very sensitive, and the rock did not present any risk to the students. Most smoke detectors (americium) or exit signs (tritium) emit more radiation. The responders did not understand the readings, nor were they able to correlate them to what was safe or dangerous. In another incident, a perceived threat was provided to several federal agencies stating that they would receive radioactive packages. In the next couple of days, several packages arrived in several different locations. In one case, responders misread the readings on their monitor and indicated that the package was providing readings into the **millirem** range, which would cause some concern as that would be unusual. The reading was actually **microrem** and was not much higher than the background level.

TABLE 9-1 Common Radioactive Sources

Name	Energy Emitted	Half-Life	Use	Decays to Form
Polonium-214	α	164 microseconds	Dust-removing brushes	Lead-210
Lead-214	β and γ	27 min	Industrial	Bismuth-214
Cobalt-60	β and γ	5.3 years	X-ray machines, industrial applications	Nickel-60
Cesium-137	β	30 years	Medical field, industrial imaging, check source	Barium-137
Radon-222	α	3.8 days	Naturally occurring gas	Polonium-218
Uranium-238	α	4.5 billion years	Ore	Thorium-234
Plutonium-239	α	24,100 years	Nuclear weapons	Uranium-235
Americium-241	α	433 years	Smoke detectors	Neptunium-237

TABLE 9-2 Common Radionuclides

Name	Alpha (α)	Beta (β)	Gamma (γ)
Americium-241	√		
Cesium-137		√	√
Cobalt-60		√	√
Iodine-129 and -131		√	√
Plutonium-239	√	√	√
Radium	√		√
Radon-222	√		
Strontium-90		√	
Technetium-99		√	√
Thorium	√		√
Tritium		√	
Uranium-238	√		√

© Jones & Bartlett Learning

material may become stable after one or two decays, but others may take many decay cycles. If a **radioisotope** decays by the emission of an alpha or beta particle, the number of protons in the nucleus is changed, and the radioisotope becomes a different element. Uranium is the base for the development of **radon**, a common radioactive gas found in homes. Eventually the radon decays into lead (**FIGURE 9-3**). Each radioisotope has a constant **half-life**, which is the amount of time for half of a radioactive source to decay (**FIGURE 9-4**). The **activity** of a source of radioactive material is a measure of the number of decays per second that occur within it. Source activity is measured in **becquerels (Bq)** or **curies (Ci)**.

One becquerel is equal to one radioactive decay per second. One curie is equal to 37 gigabecquerels (GBq), or 37 billion disintegrations per second (in the same way that 1 km is equal to 1000 m); it is also the activity of exactly 1 g of radium-222. The physical size of a radioactive source is not an indicator of radioactive strength or activity.

LISTEN UP!

Radiation detectors for emergency responders are divided into two major groups; one measures exposure to radiation, and the other measures the current amount of radiation in the area.

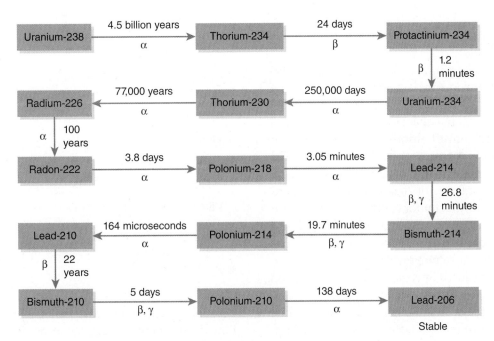

FIGURE 9-3 Chart shows the half-life for uranium-238 and the materials that it decays into. Uranium-238 eventually becomes lead-206 after billions of years. In some cases the half-life is measured in years and in some cases seconds.
© Jones & Bartlett Learning

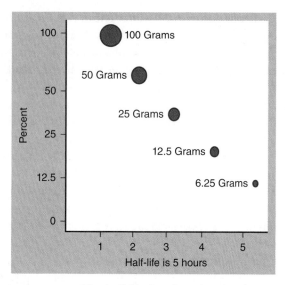

FIGURE 9-4 The half-life for plutonium is 5 hours, and so every 5 hours it is reduced 50 percent.
© Jones & Bartlett Learning

Radiation Dose

Equally as confusing as the makeup of radioactive materials is the calculation of a radiation dose. Three measurements can be used to describe a radiation dose: **Absorbed dose**, equivalent dose, and effective dose. The absorbed dose is a measure of energy transferred to a material by radiation. The absorbed dose is measured in units called **gray (Gy)** or rad (**radiation absorbed dose**). One gray is equal to 1 joule (J) of energy absorbed by 1 kg of material, and it is the System International (SI) unit for absorbed dose. One gray is also equal to 100 rad, an older unit.

To understand the impact of radiation on humans, we need to explain dose measurements and the effect of absorbed energy on the body. A **quality factor** is used to determine the potential biological damage of radiation on humans and convert absorbed dose to equivalent dose, as shown in **TABLE 9-3**. For SI units, a quality factor of 1 is assigned to beta, gamma, and **X-ray** radiation, while a factor of 20 is assigned to alpha radiation. The quality factor for neutron radiation varies depending on the energy of the neutron. The basic unit of equivalent dose in the United States is the **roentgen** (R or r), which is a value provided for the amount of ionization in air caused by X-rays or gamma radiation. Using the roentgen and applying the quality factor, we can determine the dose and impact on humans, which is known as **roentgen equivalent man** or REM (R). One roentgen equals 1 REM, and they are used interchangeably. The rest of the world uses **sieverts (Sv)** to describe the equivalent dose to man. To further define how the dose impacts humans, another factor can be added, known as a weighting factor. Target organs have a higher factor than bone, for example.

TABLE 9-3 Radiation Equivalents and Conversions

Rad	gray (Gy) or absorbed dose (AD)
REM	sievert (Sv) or dose equivalent (DE)
1 μR	0.01 μSv
100 μR	1 μSv
1 mR	10 μSv
100 mR	1 mSv
1 REM	10 mSv
100 REM	1 Sv
curie (Ci)	becquerel (Bq)
27 picocuries	1 disintegration/sec (pCi) (d/sec) or 1 becquerel
1 pCi	37 mBq
1 μCi	37 kBq
27 μCi	1 MBq
1 mCi	37 MBq
27 mCi	1 GBq
1 Ci	37 GBq
27 Ci	1 TBq
1 pCi	2.22 dpm
1 Bq	60 dpm
1 dpm	0.45 pCi

© Jones & Bartlett Learning

TABLE 9-4 Health Effects

Exposure (REM)	Exposure (Sieverts)	Health Effect	Time to Onset
5–10	0.05–0.1	Changes in blood chemistry	
50	0.5	Nausea	Hours
50–100	0.5–1	Headache	
55	0.55	Fatigue	
70	0.7	Vomiting	
75	0.75	Hair loss	2–3 weeks
90	0.9	Diarrhea	
100	1	Hemorrhage	
100–200	1–2	Mild radiation poisoning	Within hours
200–300	2–3	Severe radiation poisoning	Within hours
400–600	4–6	Acute radiation poisoning, 60% fatal	Within an hour
600–1000	6–10	Destruction of intestinal lining, internal bleeding, and death	15 minutes
1000–5000	10–50	Damage to central nervous system, loss of consciousness, and death	Less than 10 minutes
5000–8000	50–80	Immediate disorientation and coma, death	Under a minute
More than 8000	More than 80	Immediate death	Immediate

Source: From EPA, DOD and other sources.

Radiation monitors measure in three scales: Roentgen (R) or sieverts (Sv), millirem (mR) or millisieverts (mSv), and microrem (μR) or microsieverts (μSv). The dose of radiation is expressed by a time factor, typically an hour. The conversions for these values are provided in **TABLE 9-4**.

LISTEN UP!

Alpha, beta, gamma, and X-rays are forms of ionizing radiation.

Alpha (α)

Alpha radiation consists of two neutrons and two protons, which carry a +2 charge, and it is identical to a helium nucleus. Alpha is a particle and only can move a few feet in air. Its primary hazard is through inhalation or ingestion. When monitoring for alpha radiation, responders need to monitor very slowly as it does not have much energy. Scanning a person's body for alpha radiation should take several minutes. Street clothing and other personal protective equipment (PPE) provide ample protection against alpha radiation.

TYPES OF RADIATION

Radiation is comprised of two basic categories: **Ionizing radiation** and **nonionizing radiation**. Some examples of nonionizing radiation include radio waves, microwaves, infrared, visible light, and ultraviolet light. Alpha, beta, gamma, and X-rays are forms of ionizing radiation, and some characteristics of each are provided in **FIGURE 9-5**. There are two subcategories of ionizing radiation, one with energy and weight and the other comprising just energy. Alpha and beta are known as particles and have weight and energy. Gamma radiation has just energy and no weight.

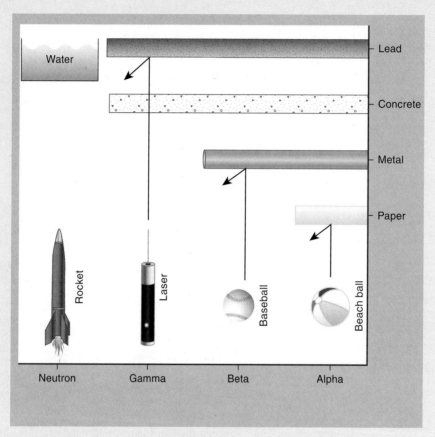

FIGURE 9-5 Examples of the various strengths of radioactive energy sources.
© Jones & Bartlett Learning

Beta (β)

Beta particles are electrons (−) or positrons (+) and weigh 1/1836 the weight of a proton. Beta radiation is of two forms, low and high energy, but both are particles.

The low-energy form is comparable to alpha but can move a little farther in air and cause more damage. High-energy beta radiation moves even farther and can cause greater harm. Beta radiation can move several feet, and higher levels of PPE are required, but turnout gear offers some protection.

Gamma (γ)

Gamma rays come from the energy changes in the nucleus of an atom, as do alpha, beta, and neutron emissions. Gamma is not a particle but is airborne energy described as wavelike radiation that comes from within the source. These waves are sometimes called electromagnetic waves or electromagnetic radiation. Gamma radiation can move a considerable distance. Lead offers some protection. It is high energy and can cause internal damage to the body without causing external damage.

Neutron (n)

Neutron radiation is not common, but it is the basis for nuclear power plants and nuclear weapons (**FIGURE 9-6**). Neutrons are ejected from the nuclei of atoms during fission when an atom splits into two smaller atoms. When fission occurs, there is a release of energy in the form of heat, gamma radiation, and several neutrons. Fission can be spontaneous, as in the case of californium-252, or induced, as in the case of uranium-235 in nuclear

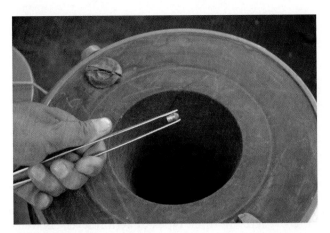

FIGURE 9-6 Californium-252, which is a neutron source; the container below is a boron pig designed for neutron sources.
Courtesy of Christopher Hawley.

FIGURE 9-7 X-ray emitting device used by bomb squads.
Courtesy of Christopher Hawley.

power plants. Isotopes that can be induced to fission are called **fissile**.

Neutrons can travel great distances, but they are stopped by several feet of water, concrete, or hydrogen-rich materials. Neutron radiation is dangerous because it readily transfers its energy to water, and the human body is more than 75 percent water. One unusual aspect of neutron radiation is its ability to activate nonradioactive isotopes, that is, make them radioactive. The nucleus of the targeted atom can capture a neutron, which makes it a different isotope. If the nucleus of the new isotope is unstable, the isotope is radioactive. For example, when natural gold, gold-197, is bombarded with neutrons, it can be activated and become gold-198, which emits beta and gamma radiation. Activation is the reason that the materials used in a nuclear reactor become radioactive and must be disposed of as nuclear waste. Activation is accomplished only by neutron radiation.

X-rays

X-rays are comparable to gamma radiation waves, as they are wavelike, but X-rays are only dangerous when the X-ray machine is energized (**FIGURE 9-7**). The radiation from an X-ray is caused when speeding electrons strike other electrons in a target. When they collide, the amount of energy created emits powerful waves of radiation. Under normal circumstances, this collision occurs in an X-ray tube and is focused on a target, usually a photographic plate of film. When the energy hits an object between the source and the photo film, it interacts with it. If it strikes an electron, it rips the electron from the atom; the wave then has weight and moves on to the target. The denser the object, the more waves are absorbed. For example, bone is denser than the surrounding tissue, so it allows fewer X-rays to

reach the film, resulting in a negative image. Therefore, on the film, the bone is light, and the flesh is dark. This ripping causes a chemical reaction, which is why X-rays and gamma radiation are much more penetrating than alpha or beta radiation.

Radiation Protection

Radiation doses should be kept "as low as reasonably achievable." This phrase is known in the nuclear industry as **ALARA** and is the cornerstone of radiation safety. The three factors that can influence radiation dose are time, distance, and shielding, concepts that should be applied to all types of chemical exposures as well. To minimize dose, stay near the radiation source for as little time as possible, stand as far away from the source as possible, and place as much shielding as possible between people and the source. Table 9-4 lists sample exposure levels, related health effects, and times to the onset of symptoms.

We need to understand this concept, as the meters generally provide measurements based on a timed exposure. Time is important, as in many cases humans can sustain a short exposure to a radiation source without being harmed. If you want to limit your radiation dose to 1 millirem per hour (mR) and the source is reading 60 mR/hr, you could be near the source for 1 minute, and your dose would be 1 mR. Being near the source for an hour results in an exposure of 60 mR. Distance from the source also plays a factor, and the safety factor with distance is provided in the fact that exposure levels are based on an inverse square law. If the source has a radiation level of 20 mR/hr at 2 feet, then moving back to 4 feet provides an exposure of 5 mR/hr. Moving to 6 feet (1.8 m) would result in an exposure of 2.22 mR/hr.

The radiation dose is important, but equally important is how quickly the dose is being applied. The measurement of this speed is the dose rate, usually expressed in an hour. The formula is Dose = Dose rate × time. As an example, suppose you are at an event where the radiation meter is reading 1 R/hr; how long would it take to receive a dose of 400 mR?

$$1\ \text{R/hr} = 1000\ \text{mR/hr}$$
$$400\ \text{mR} = 1000\ \text{mR/hr} \times \text{time}$$

$$\text{Dose} = \text{Dose rate} \times \text{time}$$
$$\text{time} = 400/1000 = 0.4$$

$$0.4 \times 60\ \text{min}/1\ \text{hr} = 24\ \text{min}$$

It would take 24 minutes to receive a dose of 400 mR.

Now that we have determined the various types of radiation and their measurement terms, we need to establish some action criteria. As was mentioned previously, time, distance, and shielding must be considered. The action levels provided in **TABLE 9-5** are assuming that the person is not wearing any protection and that there is a whole-body exposure unless noted. There are three types of action levels as well; one is for the general population, one for emergency responders, and one for nuclear industry workers. There is always a risk when dealing with any chemical substance. Think of these action levels much as you think of permissible exposure limits (PELs) and threshold limit value (TLVs).

TABLE 9-5 Radiation Action Levels

Dose	Activity Type	Note
Action Levels		
1 mR/hr (10 μSv)	Isolation zone (public protection level)	Recommended exposure limit for normal activities
5 R (50 mSv)	Emergency response	For all activities
10 R (100 mSv)	Emergency response	Protecting valuable property
25 R (250 mSv)	Emergency response	Lifesaving or protection of large populations
>25 R (>250 mSv)	Emergency response	Lifesaving or protection of large populations. Only on a voluntary basis for persons who are aware of the risks involved.
Assorted Doses		
Average Exposures to Radiation	**Cause**	**Note**
0.01 R (10 mR or 0.1 mSv) annually	Chest X-ray	Per-year exposure
0.2 R (200 mR or 2 mSv) annually	Radon in the home	Per-year exposure
0.081 R (81 mR or 810 μSv) annually	Living at high elevations (Denver)	Per-year exposure
1.4 R (14 mSv)	Gastrointestinal series medical procedure	

Source: National Fire Academy, Emergency Response to Terrorism Tactical Considerations HazMat Student Manual, February 2000, p. SM 3–25.

Radiation Monitors

Current radiation monitors provide two methods of measuring radiation: REM and count per minute (CPM), with the REM being the most important to emergency responders. A variety of detection devices are available, but the important consideration is the probe that is attached to the unit. The probe determines what type of radiation can be detected by the detector. Some monitors have a probe that detects alpha and beta radiation and a probe that measures gamma radiation (**FIGURE 9-8**). A variety of probes are available, but the most common is a pancake probe that is useful for alpha radiation.

When the amount of radiation becomes higher, the responder then switches to the internal probe, which is designed for higher levels of radiation. When turned on, a radiation monitor picks up naturally occurring background radiation, which should be in microrem or millisievert. The amount of background radiation varies from city to city and even varies within a few miles. It is important to do some testing to determine the normal levels in various parts of your community. Then responders can know when they are being exposed to radiation levels higher than background.

Types of Radiation Detectors

Many types of radiation detection devices are available to responders today, from personal radiation detectors and electronic personal **dosimeters** to survey meters and radioisotope identifiers (**FIGURE 9-9**). There are three basic sensing technologies used in radiation detection devices: **Geiger-Mueller (GM)** tube,

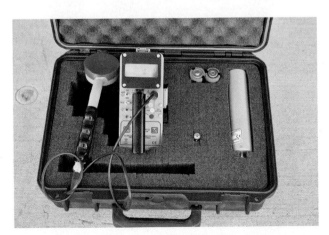

FIGURE 9-8 A typical radiation monitor that has a pancake probe for alpha and beta.
© Jones & Bartlett Learning. Photographed by Glenn E. Ellman.

FIGURE 9-9 Radiation detection device that can provide dose rate and counts per minute.
Courtesy of Christopher Hawley.

scintillation crystals, and semiconductor sensors. The type of sensor determines which form of radiation can be detected. For example, beta is best detected using a scintillation detector, while a GM tube is best for detecting gamma and beta radiation. Responders should have several different types of technology available to them.

Geiger-Mueller Tubes

GM tubes can detect alpha, beta, and gamma radiation. A GM tube consists of a gas-filled tube with an anode running down the center (**FIGURE 9-10**). An electrical current is applied to the tube, and when radiation interacts with the walls of the tube or the gas in it, electrons are freed from their atoms and flow to the anode in the center of the tube, inducing an electric current. This current is measured in electrical "pulses" and is used to determine when and how much radiation entered the GM tube. Alpha and beta radiation cause direct ionization in the chamber, while gamma and X-rays cause an indirect ionization by a reaction with the metal walls of the tube.

Geiger-Mueller detectors are generally coupled with a proportional counter, which can count the electrical pulses. The user has a choice of measuring the radiation dose or counts per minute. The GM measures and reports dose, and the proportional counter reports counts.

FIGURE 9-10 GM tubes can detect alpha, beta, and gamma radiation. A GM tube consists of a gas-filled tube with an anode running down the center; in this case the GM tube is the silver tube at the middle of the photograph.
Courtesy of Christopher Hawley.

FIGURE 9-11 Radiation pager devices typically use scintillation as their detection method. The scintillation tube is the gray tube in the front of the device.
Courtesy of Christopher Hawley.

Scintillation Crystals

A scintillation detector uses a crystal that emits visible light when hit by radiation (**FIGURE 9-11**). The most common crystal used is sodium iodide (NaI). When radiation hits the crystal, a pulse of light is produced, which is detected and amplified by a photomultiplier tube. The photomultiplier tube produces an electrical signal, which is measured to determine the amount of radiation that hit the crystal. Depending on the conditions, scintillation detectors are best for gamma detection, and they can detect very low levels of radiation. **Radiation pager** devices typically use scintillation as their detection method, as they are very sensitive and fast responding.

Gamma Spectroscopy (Radioisotope Identifier)

Some radiation monitors can identify the type of radiation source present (**FIGURE 9-12**). Each radioisotope gives off a distinct number and type of radioactive emission. Each of these emissions has a distinct, constant energy and can be used to identify the radioisotope present. The energy of the radiation can be determined with a scintillation detector and a multichannel analyzer. A spectrum is developed from the energy information. The spectrum is then compared to reference spectra stored in the onboard computer, and the radioisotope is identified. Many detection devices, such the Berkeley Nucleonics Model 940 device, can detect hundreds of radiation isotopes in a second. The device also can detect neutron sources. The Model 940 is fairly large at $12 \times 4 \times 5$ inches ($30.5 \times 10.6 \times 12.7$ cm), and it weighs 4.5 pounds (2.3 kg). Berkeley Nucleonics has a personal gamma and neutron radiation detector (the PM1703GN), as shown in Figure 9-11, that fits in the palm of your hand, weighs less than 10 ounces (0.3 kg), and can send and receive data just like a cell phone. It also can determine the identity of a radiation source in addition to determining dose or count.

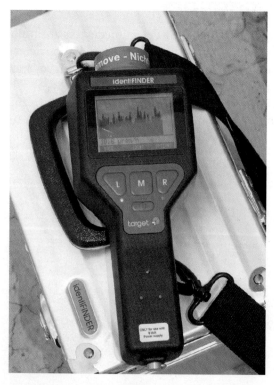

FIGURE 9-12 This radiation detector can identify the type of radiation source present.
Courtesy of Christopher Hawley.

Radiation Pagers

Another method of radiation detection is the use of a radiation pager or a dosimeter (**FIGURES 9-13** and **9-14**). These are two separate types of devices, but they are commonly used for the same purpose. The radiation pager is used to ensure the safety of personnel operating in environments where radiation exposure is possible. Radiation pagers detect X-rays and gamma, but they also detect high-energy beta radiation. Some devices can detect the presence of neutron radiation, which may be a radiation source that is considered special nuclear material (SNM) (**FIGURE 9-15**). When the pagers are turned on, they calibrate themselves to the background radiation. If a responder enters an area that has radiation levels

FIGURE 9-15 A radiation pager that can detect gamma and neutron radiation.
Courtesy of Christopher Hawley.

FIGURE 9-13 A radiation pager that can detect gamma radiation and provides a dose rate.
Courtesy of Christopher Hawley.

FIGURE 9-14 A radiation pager that only indicates levels above background in a range of 1–10.
Courtesy of Christopher Hawley.

above background, the pager will alert through either an audible alarm or vibration. It provides a reading of 1 to 10 times above the background level, which can be correlated to a range of radiation, based on the initial background radiation. It is very important, especially when dealing with terrorism or criminal events, that you check the level of background radiation when you turn on the pager. Someone intent on harming responders could place the radioactive source at the area where crews would arrive and set up. When they turned on the pagers, they would set themselves up, as the pagers would record a high level as normal background. The other form of protection is a dosimeter, which records the dose that your body is taking. Older dosimeters just recorded the dose, and the user had to visually check the dosimeter to see the exposure. New dosimeters are electronic and exist as pagers. They not only record the dose but also measure the amount of radiation that exists. They also alarm when the user enters certain levels of radiation. These dosimeters provide protection for X-ray, gamma, and beta radiation.

Through the years, radiation has been forgotten and has not been a priority for sampling. With hazardous materials teams facing terrorism concerns and high-tech criminal events, there is a need to check for radiation hazards more frequently. According to news reports and FBI records, radiation events are on the increase. According to the Nuclear Regulatory Commission (NRC), several radioactive sources are stolen each year, and radiation sources have been used in attempts to harm individuals. The most common theft is of ground imaging devices, which in some cases have significant radioactive sources that can cause harm to a community. The records of the NRC indicate that there are about 9000 devices missing in this country. When dealing with unidentified materials, packages, containers, drum dumps, and other possible criminal activities, it is very important to check for radiation. The simplest form of protection is to use a radiation monitor, establish a background level, and turn on the radiation pager. Crews should get accustomed to using these devices on a regular basis to become comfortable with the them to learn where radioactive substances exist. One warning, though, is that these devices are sensitive enough to detect radiological pharmaceuticals, such as those used in chemotherapy, so they may alert if near someone who recently had chemotherapy.

SAFETY TIP

When dealing with terrorism or criminal events, it is very important that you check the level of background radiation to make sure you are in a safe place when you turn on the pager.

LISTEN UP!

These devices are sensitive enough to detect radiological pharmaceuticals such as those used in chemotherapy, so they may alert if near someone who recently had chemotherapy.

TLD and Film Badges

For some emergency responders, radiation monitoring is done with a **thermoluminescent dosimeter (TLD)**. There are two major types of TLDs: A **film badge** and a crystal-based device. Both devices are designed to be worn by employees to monitor their exposure to radiation. They are commonly worn in nuclear facilities, hospitals, locations where X-rays are taken, and any other places involving potential exposure to radiation. The film badge TLD has a small piece of X-ray film inside a plastic case and a series of filters that assist in identifying the dose received. The filters in the Global Dosimetry Solutions Film Badge are made of aluminum, copper, lead/tin, plastic, and an open window. After a period of time, the badge is sent back to the manufacturer, who reads the exposure to the TLD and calculates the dose the wearer received.

The crystal-based TLD is a little more sophisticated than the film badge. There are a variety of crystals used, but the two common ones are calcium fluoride and lithium fluoride. When radiation hits the crystals, they become excited, and the energy inside them increases. This energy remains in the crystal until it is heated. The reading device for the TLDs heats the crystal, which then releases a photon. This photon (light) can be read, and the number of photons relates to the dose of radiation received.

After-Action REVIEW

IN SUMMARY

Hazardous materials responders need to become familiar with radiation detection, as the possibility exists for future events. Many radioactive substances exist, and any time responders are dealing with unidentified materials, they should check for radiation. Knowing how to monitor for radiation is as important as knowing what the action levels and measuring ranges are for the various types.

KEY TERMS

Absorbed dose Measure of energy transferred to a material by radiation.

Activity Measure of the number of decays per second in a radioactive source.

ALARA Radiation exposure reduction that means "as low as reasonably achievable."

Alpha Type of radioactive particle.

Becquerel (Bq) Unit of measurement for the activity level of a radiation source.

Beta Type of radioactive particle.

Curie (Ci) Unit of measurement for the activity level of a radiation source.

Dosimeter Device that measure the body's dose of radiation.

Film badge Small piece of X-ray film in a plastic case with filters to identify radiation doses received.

Fissile Descriptor of an isotope that can be induced to fission.

Gamma Form of radioactive energy.

Geiger-Muller (GM) Detection device type for low-energy materials.

Gray (Gy) Measurement of radioactivity, equivalent to RAD.

Half-life Amount of time for half of a radioactive source to decay.

Ionizing radiation Radiation that has enough energy to break chemical bonds and create ions. Examples include X-rays, gamma radiation, and beta particles.

Microrem (μR) Measurement of radiation; normal background radiation is usually in microrems.

Millirem (mR) Higher amount (1000 times) of radiation than microrem.

Neutron Form of ionizing radiation.

Nonionizing radiation Radiation that does not have enough energy to create charged particles, such as radio waves, microwaves, infrared light, visible light, and ultraviolet light.

Nuclear detonation Device that detonates through nuclear fission; the explosive power is derived from a nuclear source.

Quality factor Factor used to determine potential biological damage from radiation and to convert values from absorbed dose to equivalent dose.

Radiation absorbed dose (RAD) A quantity of radiation.

Radiation pager Detection device that signals in the presence of gamma and X-rays.

Radioactive decay Emission of radioactive energy.

Radioisotope Isotope whose nucleus is unstable and that emits radioactivity to regain stability.

Radiological dispersion device (RDD) Explosive device that spreads a radioactive material. The explosive power is derived from a non-nuclear source, such as a pipe bomb, which is attached to a radioactive substance.

Radon Common radioactive gas found in homes.

Roentgen Basic unit of measurement for radiation.

Roentgen equivalent man (REM) Method of measuring radiation dose.

Scintillation Radiation detection technology.

Sievert (Sv) Unit of measurement equivalent to REM.

Thermoluminescent dosimeter (TLD) Radiation monitoring device worn by responders.

X-ray A form of radiation much like light and gamma radiation but that bombards a target with electrons, which makes it very penetrating.

Review Questions

1. What are the four types of ionizing radiation?
2. What is the most dangerous type of radiation?
3. What types of radiation involve particles?
4. What is ALARA?
5. Which type of detector can identify the kind of radiation source or isotope?

6. What type of radiation detector technology does a radiation pager use?
7. At what level is average background?
8. What are the three levels of radiation exposure? List them from lowest to highest.

CHAPTER 10

Other Detection Devices

OUTLINE

LEARNING OBJECTIVES

Upon completion of this chapter, you should be able to:

- Identify other types of advanced detection devices that are common in emergency response
- Describe how vibrational spectroscopy and infrared spectroscopy identify chemicals
- Describe test strips and test kits that are available to emergency responders and their use in emergency response
- Describe the process of drum sampling

Hazardous Materials Alarm

Hot Gasoline—A gasoline tank truck struck the underside of a beltway bridge, killing the driver and releasing the contents of the two front compartments. First-arriving crews did a great job of protecting the bridge and setting up foam lines to extinguish the tanker fire. Then the gasoline had to be transferred from the three remaining compartments. This was an interesting undertaking because the gasoline was still very hot from the intense fire. During the transfer, the transfer hoses were hot to touch even with gloves on. Air monitoring was essential to track the vapors being expelled by the transfer operation. An infrared thermometer would provide the temperature of the hoses and the tank truck. True to form, the vapors stayed low and followed the natural lay of the land. Large isolation areas were set up to keep personnel out of the danger areas. The use of foam is always tricky because there must be a balance between vapor suppression and making the environment safe to work in. The more foam there is and the more places it is applied, the more the work areas become slippery and difficult to work in. Air monitoring can assist in determining when foam needs to be reapplied. It is best to use a photoionization detector (PID), described in Chapter 7, to determine when the foam blanket is no longer being effective. When you see the PID readings start to increase, the flammable vapors are breaking through the foam, and the foam blanket should be reapplied. The PID is the preferred detection device to determine if vapors are present because it is more sensitive than a lower explosive limit (LEL) sensor. The LEL sensor is also slower to respond than a PID, and some foam solutions can damage the LEL sensor. If you use an LEL sensor to check the foam blanket, bump test the sensor if you have the capability, and calibrate it after this use to ensure its accuracy.

1. When discussing chemical properties of a material, what is the normal temperature?

2. What other situations could an infrared thermometer be useful?

Introduction

This chapter discusses a variety of detection devices, many of which have very specialized uses. Once a substance is characterized as a fire, corrosive, toxic, and/or radiation risk, further identification needs to be made. The use of these other detection devices is integral to this identification. There are several options when it comes to identification. Some are inexpensive, while others are in the expensive category, with some over $50,000. Some are easier to use than others. The device is considered easy to use if, after basic instruction, responders can operate it and make basic determinations on the identity of the material (**FIGURE 10-1**). If responders cannot use the device, or if it is too difficult to operate, then the device may not be best suited for street-level emergency response. When a device is difficult to operate, usually only a couple of hazardous materials techs on each team can really operate the device and understand the results. Typically, these folks are off duty when they are needed the most, and others on the team can operate the device but may struggle to interpret the results. When dealing

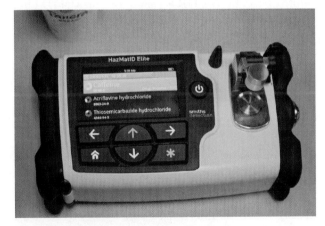

FIGURE 10-1 This is an example FTIR device from Smiths Detection.
Courtesy of Smiths Detection and Chris Hawley.

with some of the more expensive or unusual detection devices, regionalization is a good solution. Not every hazardous materials team requires one, but knowing what it can do for you and where to get one is key to ensuring a safe response.

Vibrational Spectroscopy

In emergency response, vibrational spectroscopy consists of two major technologies that are becoming very popular: **Fourier transform infrared (FTIR)** analysis and **Raman spectroscopy**. Both are similar in how they develop the resulting spectra, or "fingerprints," that help to identify a substance. They are based on the theory that all molecules are in motion and vibrate. One might think that one device is all you need, but in reality, they are both excellent tools that are complementary technologies to each other. The FTIR units were very useful in identifying unknown substances during the mysterious white powder events that occurred after the October 2001 anthrax attacks. The Raman units can also assist with future white powder events, and the combination of the two makes for an effective combination in your detection toolbox.

Fourier Transform Infrared Spectroscopy

One of the most innovative street-level detection devices in recent times has been the Fourier transform infrared (FTIR) spectroscopy devices. There are several on the market, and you should evaluate several before you select one. The Thermo Scientific TruDefender FTX, shown in **FIGURE 10-2**, is designed for the identification of solids, slurries, and liquids. Smiths Detection's FTIR is called the HazMat ID Elite (**FIGURE 10-3**). The companion to the HazMat ID Elite is the GasID designed for gases and vapors, a device that could replace colorimetric detection tubes (**FIGURE 10-4**). Although FTIR is very promising

FIGURE 10-3 The Gas ID sampling pump, which draws the sample into the stainless steel container, which is then inserted into the GasID.
Courtesy of Smiths Detection and Chris Hawley.

FIGURE 10-4 The GasID device, which is an FTIR, which samples the gas from the gas pump.
Courtesy of Christopher Hawley.

and a good identification technology, there are some drawbacks, as with all detection devices. FTIR devices cannot detect every substance. Having some understanding of chemistry helps a responder determine which materials FTIR is useful for and which it does not identify. The type of molecular bond determines whether the FTIR can identify the material. There are two types of bonding: Ionic and covalent. All chemicals bond in one of these manners. In ionic bonding, there is an exchange of electrons; one atom loses an electron to the atom it is bonding with. In a covalent bond, the electrons are shared by both atoms. When atoms share the electrons, they exhibit motion, or molecular vibration. This vibration is specific to each type of bond. To perform its analysis, the FTIR sends infrared radiation, which is a collection of waves of varying frequencies, to the sample material. The motion in the covalently

FIGURE 10-2 An example FTIR device from Thermo Scientific.
Provided by Thermo Scientific.

bonded atoms generates similar waves with the same frequencies. When the infrared radiation hits the atoms, the bonds absorb the radiation, which results in peaks being formed in the infrared spectrum. The resulting pattern of peaks are compared to the internal reference library that has thousands of chemical entries. The closer the match of peaks (also called spectra [plural] or spectrum [singular]) that a peak is, the higher the confidence in a match. The higher the confidence is, the higher the reported correlation is by the instrument. Smiths recommends looking at the spectra to check for accuracy. The FTIR can detect common industrial materials, chemical warfare agents, organic solvents, explosives, pesticides, drugs, and ionic compounds.

The library for the HazMat ID includes over 25,470 chemicals, and a big advantage of this technology is that users may add spectra of other materials they encounter. The GasID has a library of 5500 materials and has the same capabilities as the HazMat ID to add spectra. Both Smiths and Thermo offer a reach-back program. If the responder runs a sample and the FTIR does not provide an identification, the user can contact the manufacturer. The reach-back service looks at the spectra to try to identify the material. The FTIR cannot detect pure metals and nonmetal elements, such as lead and arsenic. It also cannot detect table salt or diatomic gases, such as chlorine and oxygen. The FTIR recognizes water and carbon dioxide and reports the resulting spectra, which for an experienced user is not a problem. The biggest drawback of FTIR devices is that they do not recognize components of mixtures, although they do well with three compounds. If the minor component of the mixture is less than 10 percent, the material is not seen by the instrument. An experienced user can eliminate this trouble by subtracting the known spectra and examining the remaining spectra. Practice using this instrument with several different types of materials on a regular basis is highly recommended.

Raman Spectroscopy

Raman spectroscopy is also an exciting technology. It has a long history in the research and laboratory fields and is making an entrance into the emergency response arena. A Raman spectroscopy device uses a laser as its infrared light source (**FIGURES 10-5** and **10-6**). The laser, with its single frequency, hits the sample material, and some of the light is reflected and scattered, while the remaining light passes through the sample. Each molecule reacts in a different manner when hit with the laser, and the resultant scattering is unique to each type of molecule. The advantage with a traditional Raman device is that the sample can remain inside a

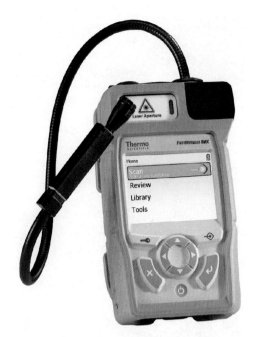

FIGURE 10-5 A Raman device from Thermo Scientific.
Provided by Thermo Scientific.

FIGURE 10-6 A Raman device from Smiths detection.
Courtesy of Smiths Detection and Chris Hawley.

glass container or plastic bag while being analyzed. The responder does not have to come in contact with the chemical to provide the analysis. However, the responder does have to come very close to the container, and in most cases, the sample material must be within several millimeters of the laser. Cobalt (Agilent) has developed a Raman device shown in **FIGURES 10-7** and **10-8**. This device has a distinctive advantage in the Raman field as it can see through multiple barriers of nonmetal material. The layers of material do not have to be

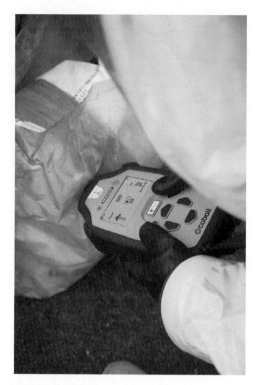

FIGURE 10-7 This is an example Raman device from Agilent (Cobalt Light).
©Agilent Technologies, Inc. 2016, 2018
Reproduced with Permission, Courtesy of Agilent Technologies, Inc.

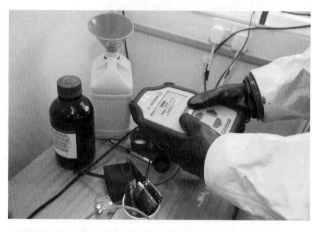

FIGURE 10-8 This Agilent (Cobalt Light) Raman can see through containers, which is a unique feature. Standard Raman devices can see through clear containers, but this device can see through nonmetal opaque containers.
©Agilent Technologies, Inc. 2016, 2018
Reproduced with Permission, Courtesy of Agilent Technologies, Inc.

clear. Their Resolve device uses spatially offset Raman spectroscopy (SORS) which enables it to see through containers. The SORS technology conducts an analysis of the container and the contents inside the container. The algorithms within the device can calculate which is the container and what is the material of interest. It can see through several millimeters of packaging, and there can be multiple layers of packaging. The sample time is about 1 minute, and the device has a comparable library to the other devices. The biggest advantage of the Raman device is also its biggest disadvantage, which is the laser. Both the Ace-ID, made by Smiths Detection, and the Thermo Scientific FirstDefender RM have two methods of performing analysis. The responder can put a sample into a 4-mL glass vial, which is then placed inside the device. The analysis is performed by a shielded laser. The other possibility is to point and shoot the laser at the sample material; in this case, the laser is unshielded. The Thermo Scientific FirstDefender RMX has an extension wand that can be used to provide some flexibility to the sampling technique. The laser used is more dangerous than a laser pointer and could cause injuries if used improperly. Laser eye protection is recommended when using an unshielded laser. The point-and-shoot method is standard for both the Smiths unit and the Thermo Scientific unit.

FTIR and Raman Technologies Compared

In addition to the ability to penetrate a clear container, the Raman units do not recognize water; this allows the user to test for contaminants in water by comparison. FTIR units "see" the water, which is a strong absorber of infrared light. To a user of FTIR with basic experience, the water spectrum does not present an overwhelming obstacle. After running a few samples on an FTIR unit, many users can recognize the water spectra and eliminate it, leaving the contaminants for analysis. Mixtures are another question. The Thermo Scientific unit has a library of 1000 pure materials, and through analysis, the responder can look at hundreds of thousands of mixtures. Raman has the same issue as the FTIR units and has trouble with mixtures. Raman cannot identify materials that are less than 10 percent of the mixture. Thermo Scientific and Smiths Detection have both taken some steps to help identify mixtures through unique spectroscopy software. In simplified terms, if the sampled material is not found in the library, the software determines whether it is made up of a mixture of materials in the library. The device reports that it has found the sample to be a mixture and provides a list of possible components. The device does not indicate the concentration of the components in the mixed materials. Another issue with Raman units is the possibility of the laser igniting thermally sensitive (reactive) materials. Dark-colored materials seem to be the most problematic. A material such as black powder, silver azide, match tips, black plastics, latex paint, and cardboard can smoke or possibly be ignited by the laser at the maximum setting. Safe work practices, such as never aiming the laser at a large amount of product, enhance user safety. Ignition is a possibility predominantly with dark reactive materials because they cannot dissipate the heat buildup produced by absorbing the laser energy. As with any testing process, it is always advisable to take a very small sample away from the larger pile of material to conduct testing. For example, a responder would not conduct a water reactivity test on a large pile of unidentified and potentially reactive material. Combining these technologies is a winning approach because FTIR and Raman complement each other. FTIR is a good detection technology for some materials that Raman is not the best at and vice versa. FTIR requires the user to obtain a sample to place onto a testing surface, while Raman can analyze materials inside a container. Raman can handle corrosives better because there is usually water in corrosive materials, but FTIR is better for alcohols. The FTIR can detect HCl (purely ionic when dissolved in water) because hydrogen bonds with water, and the altered water spectrum uniquely identifies the muriatic acid. Raman cannot see muriatic acid. Materials that fluoresce are problems for Raman but do not affect FTIR. For example, Raman cannot recognize cyclosarin, a military chemical agent that is extremely toxic, because it fluoresces when hit with a laser. The FTIR unit sees it very well without any fluorescent effect.

Infrared Spectroscopy

A detection technology that has emergency response application is a single-beam infrared (IR) spectrometer, which is used by the MIRAN SapphIRe XL device, manufactured by Thermo Electron Corporation (among other IR devices on the market). The device can detect up to 150 different gases using infrared spectroscopy. This type of spectroscopy, also known as absorption spectroscopy, is the basic form for IR analysis by other devices, such as the IR sensor used in flammable gas detection. An infrared light source is produced and sent through a sensor housing. The housing splits the beam into two parts: One a reference beam and the other for gas analysis. The gas sample is sent into the analysis side of the sensor, along with the IR beam. On the other side of the sensor is a detection device that determines the amount of IR light reaching the sensor. It compares the amounts sent from the reference side and the analysis side. Each chemical detected by the IR spectrometer has a known amount of absorbency for IR light, which is compared to the amounts stored in the device's library. The use of this technology is primarily limited to materials with covalent bonds, typically organic materials. As with other IR detection techniques, mixtures present problems; the purer a substance is, the better the results will be.

Mercury Detector

A mercury detector is used to detect mercury vapors in the air (**FIGURE 10-9**). One unit has a measuring range of 0.001–0.999 mg/m^3 of mercury, and Chapter 1 discusses mg/m^3 conversion to parts per million. The unit has a sensor lined with gold foil, which attracts mercury. If the unit is placed in an environment that has mercury vapor in the air, the mercury will be attracted to the gold foil. As the mercury collects on the gold foil, the electrical resistance changes and causes a corresponding reading on the display. This type of detector is very susceptible to false readings if the instrument is hand carried or is moving around.

The best situation is to have the instrument placed on a table or other stable environment and bring the potential contaminant up to the instrument. Another method is to take three readings and average the readings to arrive at a value that should be close to the actual level. When dealing with possible victim contamination, it is best to remove his or her clothes, bag them, and then later insert the probe of the instrument into the bag to sample the atmosphere. Hazardous materials responders do not encounter mercury on a regular basis, but since it is a chronic health hazard and a very toxic material through chronic skin contact, a detection method should be available, and you should have an emergency contact to obtain this type of device. A low-cost alternative is colorimetric tubes that can detect mercury vapors. These tubes have a detection range of 0.05–2 mg/m^3 for Dräger and 0.05–2.5 mg/m^3 for Sensidyne, which is more than adequate for sampling purposes. In many cases, the material is spilled in a home, and a concern for the health and welfare of the residents needs to be a high priority. Some terrorists use mercury in explosive devices, so mercury may be present at a terrorism incident.

FIGURE 10-9 A mercury detector with a gold foil sensor. Mercury has an affinity for gold and is attracted to the sensor.
Courtesy of Christopher Hawley.

A 911 call was received for a mercury spill, in which a barometer had been broken, and the mercury had spilled out. Upon arrival at a local retirement community, a moving crew was noted, and the people who had called 911 were located. It was determined that a large barometer had been broken during the move, and when it was unpacked, mercury had spilled out in the resident's apartment (**FIGURE 10-10**). The apartment was the last apartment on the other side of the building, a considerable distance away from the front of the truck. It was possible that mercury was spilled throughout the hallways leading out of the building, and since this pathway led to other areas of the building, the mercury could have been throughout the building. The release of mercury even in small quantities is a big concern; it is not an acute health hazard but is a serious chronic health hazard. The Agency for Toxic Substances and Diseases Registration (ATSDR) has a website located at https://www.atsdr.cdc.gov/mercury/ that has considerable information. There is not a need for extensive personal protective equipment (PPE); just avoid contact. The inhalation of vapors is a long-term issue, not a short-term problem. Isolation of the contaminated areas is essential to make sure the mercury is not tracked elsewhere. The persons who were in or near the spill were isolated, and those who it was suspected got the mercury on them were told to remove their clothes and were provided other clothes. The clothes were bagged and tagged and placed outside in the sun (**FIGURES 10-11** and **10-12**). At the time, the Maryland Department of Environment (MDE) had the only mercury detector in the state, so it was requested for assistance. Complicating this was the fact that about 10 people had been in and out of the apartment looking at the spill. One person who likely had mercury on him had driven to a local store to get some food, which

FIGURE 10-11 Using a magnifying glass looking for mercury. A flashlight shined at a right angle assists in this process.
Courtesy of Christopher Hawley.

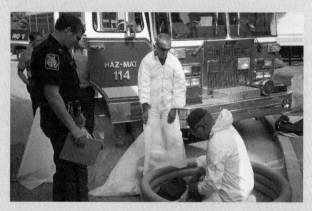

FIGURE 10-12 A police officer tracks the person's valuables while Bob Swann of the Maryland Department of Environment (MDE) checks for contamination.
Courtesy of Christopher Hawley.

had the potential for moving the problem off-site. By the time the monitor arrived, we had established a monitoring plan to check for contamination. We checked the hallways and luckily did not find any contamination. We checked the apartment where there was a visible mercury spill and found this to be the only location where there was any mercury. We then checked and found four sets of clothes that had been contaminated and four sets of shoes that were contaminated; the remaining clothes were clear. During this process, we found a worker from another building who had been in the room and had cleaned up some of the mercury and placed it in a biowaste container for disposal. Some of the tools used to clean up the spill were put in one of several dumpsters. It was also determined that the box that held the barometer had been taken back outside and placed with the other trashed boxes and packing material. All this was bagged as waste for disposal, as it could not be determined what packing material was in the contaminated box. Unfortunately, when mercury is spilled on carpet, the carpet must be picked up and destroyed, as there is no method of decontamination that would not degrade or destroy the carpet. The barometer was an antique barometer built in 1790 and was worth $30,000. The carpet, also an antique, was worth $100,000.

FIGURE 10-10 Visually looking for the mercury spill in the apartment. Once the spill is located, the mercury detector is set down and the air sampled near the spill. The mercury detector is too shock-sensitive to operate while walking.
Courtesy of Christopher Hawley.

HazCat™ Kit

The **HazCat™** kit is a system that uses a chemistry set to assist in the identification of unidentified solids, liquids, and gases (**FIGURE 10-13**). The HazCat kit uses colorimetric tubes to identify unidentified vapors. Information on the use of colorimetric tubes is provided in Chapter 8, so it is not repeated here. For many of the solids and liquids, the kit uses a flow chart to sort through the various chemical families, which would be the minimum identification. The HazCat kit does a great job of providing more specific identification. Much like the colorimetric system, the HazCat kit relies on the fact that chemicals within the same family react in a similar fashion. By using a series of chemical tests, including mixing the unidentified with several chemicals called reagents, this system uses the process of elimination to identify and/or characterize a sample. One of the beneficial tests is the iodine chip test (**FIGURE 10-14**).

Several types of HazCat kits are available, some for specific applications. For those interested, HazCat sells their entire line in one kit, which covers a wide variety of potential hazards. One of the significant issues with unknown white powder events is the fact that most responders and laboratories were concerned only with testing for anthrax. Very little testing of the white powders identified what the white powder was, nor was the associated risk exactly determined. Any of the HazCat kits will point you toward a general hazard category, if not get you close to an identification. HazCat has white powder kits, biological test kits, methamphetamine lab test kits, asbestos and lead kits, weapons of mass destruction (WMD) kits, rocket fuel kits, and several others. The system requires specific training, and a background in chemistry helps. Frequent use or training is required to maintain proficiency. The system does very well to classify unknown materials, and responders who regularly use the kit can make a classification in about 20–30 minutes, while users who are unfamiliar with the kit take much longer. In emergency response, identify the risk the material presents, make the situation safer, and then under less stressful conditions make the identification using the HazCat kit. Typically, the one big area that is deficient in effective sampling is the identification of unidentified solids. Some of the initial concerns with solids are found in **TABLE 10-1**, and some concerns with liquids are found in **TABLE 10-2**. Sugar and flour may seem unusual to you, but these two items are used often to mimic terrorism agents in terrorist hoaxes, so these tests are included.

FIGURE 10-13 HazCat kit.
© Jones & Bartlett Learning. Photographed by Glenn E. Ellman.

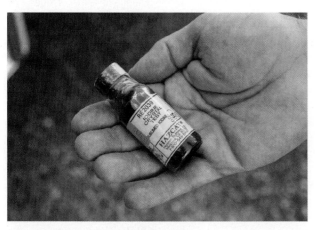

FIGURE 10-14 Iodine chip test.
© Jones & Bartlett Learning. Photographed by Glenn E. Ellman.

TABLE 10-1 Concerns with Unidentified Solid Materials
Water reactivity
Air reactivity
Explosive characteristics (oxidizers)
Peroxidized materials
Toxicity (cyanide, sulfides)
Sugar test
Flour test

© Jones & Bartlett Learning

TABLE 10-2 Concerns with Liquids
Water reactivity
Air reactivity
Peroxidized materials
Toxicity
Narrow identification to a chemical family or group

© Jones & Bartlett Learning

Provided in **TABLE 10-3** are a few of the tests that can be accomplished by the HazCat kit for solid materials. The HazCat system has full flow charts and provides extensive documentation and a process for determining the identity of unidentified solids. The tests in the table provide a quick overview of some of the initial characterizations listed in Table 10-1 and are in no way intended to replace the HazCat system. Refer to the HazCat kit for additional safety concerns and more information. **TABLE 10-4** provides just a few of the many tests that can be accomplished by the HazCat kit for liquid materials. The HazCat system has full flow

TABLE 10-3 HazCat Solid Sampling

Test	Instructions	Results
Water reactivity	Place a pea-size sample on the watch glass. With a pipette, place two to three drops of neutral water on the sample.	A lot of bubbling indicates severe water reactivity; a little bubbling indicates slight water reactivity.
Cyanide test	Add a pea-size amount of the solid material to ½ inch of water in a test tube. In another test tube, place 1¼ inch of iron citrate and a pinch of ferrous ammonium sulfite. Mix the two tubes together, pouring the second tube's contents into the first. Shake this mixture for at least a minute. Add 5–10 drops of hydrochloric acid (3N) to the mixture.	A color change to dark Prussian blue is an indicator for cyanide.
Oxidizer test	Place a pea-size sample on the watch glass, add one to two drops of hydrochloric acid (3N) to the oxidizer test paper, and then dip the paper into the sample. [Author's note: The HazCat kit instructs you to add the acid to the unidentified material for some tests after this initial test. My preference is to add the acid to the unidentified, test with the test paper, and then continue down the chart with the other HazCat tests.]	If the paper turns blackish blue, blue, purple, or black, an oxidizer is present. [Author's note: The HazCat system has several other tests to run to confirm the presence of an oxidizer. It is recommended that you continue with those tests.]
Peroxide test	Place a pea-size sample of the unidentified material in a watch glass, and add four to six drops of water so that the sample is a slurry-like solution. Dip the peroxide paper into the sample.	Wait up to 15 seconds, look for a color change from white to blue, and refer to the package for the concentration of the sample.
Sugar test	**Warning**: Be careful where you point the test tube while heating this solution. In one test tube, mix ¼ inch (6.4 mm) of copper sulfate with a ¼ inch (6.4 mm) of alkaline tartrate. Stopper and shake the mixture until it becomes a rich blue color. In another test tube filled with ½ inch (13 mm) of water, add half of a pea-sized amount of the unidentified. If you have a liquid and you want to test it for sugar, place ¼ inch (6.4 mm) of the liquid into a test tube, and follow the remaining instructions. Add 1 drop of 3N hydrochloric acid to the test tube containing the unidentified. Over a torch flame, gently heat the acidified unidentified to almost boiling. Be careful when heating liquids in a test tube, as they can easily boil out. Add equal amounts of the blue sugar test to the heated mixture of the unidentified. Try not to fill the test tube to over half full. If there is no orange to the copper precipitate, gently heat the mixture some more.	A color change to yellow/orange then to red/brown precipitate indicates sugar. If the solution starts to turn orange or yellow, then turns clear green, too much acid was added to the unidentified, or the unidentified was too acidic. Flour and starch are complex sugars and may give a slight reaction. If the unidentified forms either a suspension or a gel, perform the flour test.
Flour test	Place a pea-size amount of the unidentified in a watch glass. Add two to three drops of potassium iodide to the watch glass.	Color change to orange to blue/black indicates starch or flour.

© Jones & Bartlett Learning

TABLE 10-4 Liquid Sampling

Test	Instructions	Results
Water reactivity	Place a small amount ¼ to ½ inch (6.4–13 mm), of the unidentified liquid in a test tube. With a pipette with neutral water, place two to three drops into the test tube.	A lot of bubbling indicates severe water reactivity; a little bubbling indicates slight water reactivity.
Specific gravity/polarity	Put ¼ to ½ inch (6.4–13 mm) of water in a 40-mL (1.4 oz) vial, then add four to six drops of the sample to the vial.	If the material dissolves in the water and mixes with the water, it is polar (water soluble). If the material does not mix with the water, it is considered nonpolar. If the sample floats on the water, it has a specific gravity less than 1. If the material sinks in water, it has a specific gravity of greater than 1.
Flame test	Take the unidentified sample in a watch glass and light a match, preferably on a pair of forceps. Starting at 6–8 inches (15–20 cm) above the sample, slowly lower the match to the sample, continuing until you reach the sample. If it has not ignited, touch the match to the sample.	If the sample ignites at a distance of 2 inches (5 cm) and above, the sample is extremely flammable. If the sample ignites from just above the sample to 2 inches (5 cm), the sample is flammable. If the material ignites when the match touches the surface or is slightly above and ignites after heating a few seconds, the material is combustible. If the match increases in flame height when touching the surface, the material is slightly combustible. If the match is extinguished when it touches the surface of the material, it is noncombustible.
Oxidizer test	Place four to six drops of the sample in the watch glass and add one to two drops of hydrochloric acid (3N) to the oxidizer test paper. Then touch the paper to the unidentified. [Author's note: The HazCat kit instructs you to add the acid to the unidentified for some tests after this initial test. My preference is to add the acid to the unidentified, test with the test paper, and then continue down the chart with the other HazCat tests.]	If the paper turns blackish blue, blue, purple, or black, an oxidizer is present. It is important to watch for the results immediately as after time the paper will change color due to the oxygen in the air. [Author's note: The HazCat system has several other tests to run to confirm the presence of an oxidizer. It is recommended that you continue with those tests.]
Peroxide test	Dip peroxide test paper that has been wetted with water into the sample. For hydrocarbons, dip paper into sample and allow the liquid to evaporate. Then dip the test strip into water.	Wait up to 15 seconds and look for a color change to blue. Refer to the package for the concentration of the sample.

(continued)

TABLE 10-4 Liquid Sampling (continued)

Test	Instructions	Results
Hydrocarbon characterization (iodine chip test)	Place ½ inch (13 mm) of the sample in a test tube. Add one or two iodine chips to the test tube. Some slight agitation may be necessary. The second part of the test, after the color change, is to add some distilled water to the test tube.	Red (burgundy) indicates an unsaturated hydrocarbon. If the unidentified floats when several drops of water are added to the test tube, it is most likely benzene or toluene. If the unidentified sinks in water, it is probably a halogenated hydrocarbon. The most common halogenated hydrocarbons are perchloroethylene, trichloroethylene, and chlorobenzene. If you add water and the top is burgundy/red and the bottom is yellow or clear, then BTEX (benzene, toluene, ethylbenzene, and xylenes) and polar liquids are indicated. Purple indicates a saturated hydrocarbon.
		A purple indication is only possible when there is no functional group except a halogen on a saturated hydrocarbon (no double bonds). If the unidentified floats when water is added, it is most likely paint thinner, naphtha, mineral spirits, or Stoddard solvent. If the unidentified is slightly oily, it is kerosene. Siloxanes are rare products that float and produce a purple color change. If the unidentified sinks when water is added, it is either carbon disulfide or a high-molecular-weight hydrocarbon. Carbon disulfide is highly volatile, extremely flammable, toxic, and smells like rotten pumpkins. The most commonly encountered halogenated hydrocarbons are carbon tetrachloride and methylene chloride. If you add water and the top is purple and the bottom changes to yellow or clear, then paint thinner, mineral spirits, kerosene, or jet fuel is indicated.
		Yellow or light orange indicates polar hydrocarbons such as acetates, alcohols, ethers, esters, nitriles, acids, or a mixture.
		Orange is predominant when there are two or more characteristics. If after water is added the mixture separates and produces a new color, a mixture is indicated.
		[Author's Safety Note: If the material also has a pH of >10, dispose of the material carefully as it may be ammonia, and the addition of iodine and ammonia forms nitrogen tri-iodide, a material that may explode if it is allowed to dry and is then disturbed. It is shock sensitive. PPE is important when sampling.]
		Red/orange indicates a double bond and a polar group such as phenol.
		Brown or opaque is an indication of gasoline. Add water to the solution. The solution should separate, the bottom should be yellow, and the top will remain brown. This change also indicates the presence of gasoline. A brown color also may indicate a mixture of hydrocarbons.
		Gasoline turns brownish black and stains the sample jar in few minutes with black spots and a black ring around the liquid level. Diesel fuel also turns muddy brown, but the black spots begin to appear after 1 to 2 hours, and the ring around the liquid level does not appear for 6 to 12 hours, although the bottom of the sample jar may be slightly brown/black after a few minutes.
		Unidentified with a pH near 11 or above 11 reacts, smokes, and remains indicates amines, guanidines, pyridines, or other nitrogenated organic compounds; hydrazine salts or weak hydrazine solutions. Look for copious amounts of purple smoke.
		Unidentified with a pH near 7 reacts, smokes, and remains (or changes to dark red) indicates turpentine.
		Unidentified reacts, liquid boils and becomes clear indicates a triple bond.
		Unidentified immediately catches fire indicates hydrazine.
		No color change and the viscosity of kerosene indicate a carbon chain of 17 carbons or more.

Excerpted with permission from Robert Turkington, HazCat Abridged Manual for Field Use 1995 (HazTech Systems Inc. of San Francisco, CA).

charts and provides extensive documentation as well as a process for determining the identity of unknown liquids. Refer to the HazCat kit and instructions for more information. These tests are provided to you as they give a quick overview of some of the initial characterizations listed in Table 10-2.

SAFETY TIP

Always follow the instructions in the HazCat kit prior to performing these tests listed below.

Test Strips

One of the easiest of multiple chemical tests for street responders is the chemical test strip (**FIGURE 10-15**). Several strips are on the market, and JV Manufacturing produces two of the most common ones (on the market the longest): **Chemical classifier** (Spilfyter™) and the **wastewater** version. Recently there have been two additions to the market: The Water Safe Test Kit and the **Chameleon**. Both JV Manufacturing test strips share some duplication, and responders should have both available. Much like pH paper, these strips are for testing unidentified liquid samples and can assist in identifying unknown materials. There are five tests on each strip, and the manufacturer intends that the strip be dipped into the unidentified material. However, it is recommended that a sample be taken from the unidentified material using a pipette and placing a small drop onto each test area (**FIGURE 10-16**). Some useful interpretation can be done if the sampling is done in this way. Pure compounds initially produce more vibrant color changes, while mixtures or less concentrated chemicals produce less vibrant color changes and usually take a little longer. If the strips were placed into an unidentified chemical that was slightly thick and had color to it, a color change is more difficult to see. The Spilfyter™ tests for pH; oxidizer, fluoride, and chlorine/iodine/bromine risks; and hydrocarbons. See **TABLE 10-5**. The oxidizer risk test is important, as it is a quick test for potentially explosive materials. The test for fluoride assists in the identification of hydrofluoric acid (HF). If you get a pH of 0–1, are not sure what type of acid is present, and the test for fluorine is positive, then most likely the acid is hydrofluoric. Only a couple of fluorinated acids exist, with the most common being HF. The wastewater strip samples for pH, hydrocarbons, nitrite risk, and lead risk. One important test on the wastewater strip is the test for lead, which is commonly dumped as part of old paint waste. With these strips we can very quickly identify the chemical family that a material comes from and make some effective and safe response decisions. These test strips are a good lead-in for the tests found in Table 10-4. One good thing that these strips can do is rule out what kinds of chemicals are not present. The lead test can detect lead levels below the EPA action level of 15 parts per billion (ppb).

LISTEN UP!

One good thing that these strips can do is rule out what kinds of chemicals are not present.

Morphix Technologies provides several types of colorimetric tests. They offer a Chameleon armband, TraceX explosives kit, SafeAir badge, and a ChromAir badge. The Chameleon has several variations, and the hazardous materials detection kit can conduct up to 10 tests at a time. It is intended for vapor sampling. The Chameleon tests are designed for military applications and can be immersed in freshwater or saltwater for an hour without any degradation. The operating range is -22 to 122°F (-30 to 50°C), and, once opened, the strips

FIGURE 10-15 Spilfyter™ strips, a Chemical Classifyer and a wastewater strip, are useful in classifying unidentified substances.
Courtesy of Spilfyter™ and Chris Hawley.

can be used for up to 24 hours. One drawback is that the strips have a 2-year shelf life. They will detect chemicals at permissible exposure levels in less than 30 minutes. At half of the immediately dangerous to life or health (IDLH), they will indicate in less than 5 minutes.

Chameleon detects acids and bases, chlorine/fluorine, hydrogen fluoride, hydrazine, hydrogen sulfide, iodine, phosgene, phosphine, and sulfur dioxide. The test strips fit into an armband and require the user to watch each of the test strips (**FIGURE 10-17**).

Clan-Meth detection kit can detect ammonia, hydrogen sulfide, iodine, acids, and phosphine.

Chemical suicide detection kit detects bases, hydrogen sulfide, acids, phosphine, and sulfur dioxide.

Arson Investigation kit can detect ammonia, hydrogen cyanide, hydrogen fluoride, hydrogen sulfide, acids, nitrogen dioxide, phosgene, and sulfur dioxide.

TraceX Explosives kit detects nitroaromatics (TNT, DNT, tetryl); nitramines and nitrate esters (RDX, HMX, PETN, EGDN, NG, R-Salt, Semtex); chlorates and bromates (potassium chlorate, potassium bromate); peroxides (TATP, HMTD); acids (nitric acid, sulfuric acid); and bases (potassium hydroxide, sodium hydroxide). These tests can detect trace levels and have a 2-year shelf life (**FIGURE 10-18**).

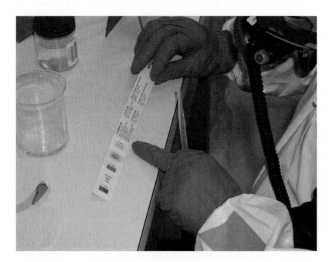

FIGURE 10-16 Although the instructions recommend that you dip the whole strip into the unidentified liquid, it is best to place a drop of the sample onto the individual tests. In this way, a closer observation of the results can be accomplished.
Courtesy of Spilfyter™ and Chris Hawley.

TABLE 10-5 Spilfyter™ Chemical Test Strips	
Test	**Sensitivity**
Chemical Classifier Strip	
pH	0–13
Oxidizer	1 mg/L hydrogen peroxide
Petroleum hydrocarbon	10 mg/L gasoline
Chlorine, bromine, and iodine	1 mg/L chlorine
Fluoride risk	20 mg/L fluoride
Wastewater Strip	
pH	0–13
Nitrites	1 ppm nitrite
Petroleum hydrocarbons	10 mg/L gasoline
Lead	20 mg/L (20 ppm) lead
Hydrogen sulfide	10 ppm

© Jones & Bartlett Learning

FIGURE 10-17 Chameleon, which is a colorimetric test kit for a variety of chemicals.
Provided by Morphix Technologies.

SafeAir badges are for individual chemicals and are available for ammonia, aromatic isocyanates, carbon monoxide, chlorine, chlorine dioxide, formaldehyde, hydrazine, hydrogen chloride, hydrogen sulfide, mercury, ozone, phosgene, and sulfur dioxide. An alert appears when in the presence of the material when at the time weighted average (TWA) range. Each badge has a varied detection level (**FIGURE 10-19**).

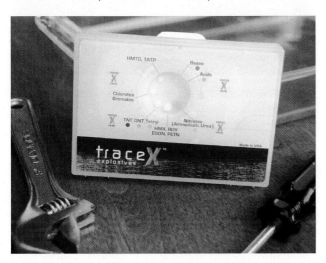

FIGURE 10-18 Trace-X is a rapid explosive test.
Provided by Morphix Technologies.

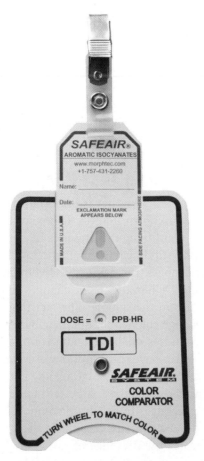

FIGURE 10-19 SafeAir badges indicate the presence of the target material.
Provided by Morphix Technologies.

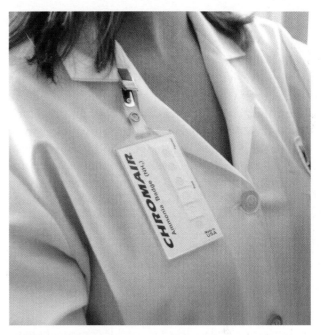

FIGURE 10-20 ChromAir provides exposure levels for target materials.
Provided by Morphix Technologies.

ChromAir chemical detection badges are single-gas badges that are available for acetone, ammonia, carbon monoxide, chlorine, formaldehyde, glutaraldehyde, mercury, methyl ethyl ketone, methyl isobutyl ketone, and ozone. These badges provide exposure levels from one-tenth to two times the permissible exposure limit (PEL) (**FIGURE 10-20**).

Other test strips can assist in the identification of unknown materials. The most common tests are for peroxides, chlorides, fluorides, and other materials. Many of these require refrigeration and are quite expensive. A section on the HazCat kit provides a description of the common test strips that are useful for emergency responders.

One test kit that has both standard hazardous materials and terrorism response use is the pesticide test ticket. The Agri-Screen Tickets produced by Neogen Corporation are extremely sensitive devices and easy to use. They can detect down to the parts per billion range for organophosphates, thiophosphates, and carbamate pesticides. Essentially any material that is a cholinesterase inhibitor will be picked up by this test, which includes the military warfare nerve agents. The only interferences are pHs less than 3 or greater than 8 or materials that use dyes (chromophoric material). The most important point to remember in using this kit is to make sure the glassware and tools are cleaned with a bleach solution to eliminate false positives. Any residual pesticide will cause the next test to be positive.

Oxidizer Test Strips

This test strip is also called potassium iodide starch paper (**FIGURE 10-21**). These strips detect strong oxidizers such as nitrites and free chlorine. Oxidizers tend to react violently and should be quickly ruled out. To use the strips, you add a drop or two of hydrochloric acid to the strip, and then touch the strip to the unknown (**FIGURE 10-22**). The best method is to use a clean tool and remove a small pea-sized amount of the unknown from the entire collection of the unknown; then you can touch the strip to the small sample. This white strip turns blue-black in the presence of an oxidizer. The reaction will be quick, generally a few seconds. The paper will turn black after a few minutes in the air once the acid was added to the strip.

Fluoride (F) Paper

This paper is used to detect a fluoride component in an unknown or detect hydrogen fluoride (**FIGURE 10-23**).

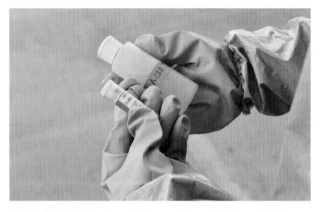

FIGURE 10-21 Oxidizer paper, which turns black in the presence of an oxidizer.
© Jones & Bartlett Learning. Photographed by Glenn E. Ellman.

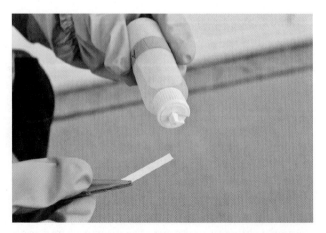

FIGURE 10-22 To activate the oxidizer paper place a drop of HCL on the paper.
© Jones & Bartlett Learning. Photographed by Glenn E. Ellman.

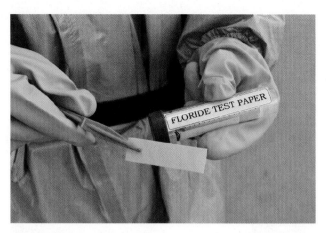

FIGURE 10-23 F paper, which indicates the presence of fluorine.
© Jones & Bartlett Learning. Photographed by Glenn E. Ellman.

When testing an unknown that is not acidic, add a small drop of an acid (hydrochloric acid) to the paper, and then add the paper to a small pea-sized sample of the unknown. When testing an acidic material, you do not have to add the acid to the paper. The F paper is normally pink and changes to a yellowish-white color in the presence of fluoride. The threshold of detection is 20 mg/L of fluoride. This paper may also react to chlorates, bromates, and sulfates in large amounts. There are methods to avoid these interferences, but they are not practical for emergency response. If you are confronted with an unknown high vapor pressure acid, the most likely acid would be hydrochloric acid. However, it could be hydrofluoric acid, which is significantly more dangerous. Using F paper will rule out the hydrofluoric acid.

Water-Finding Test Paper

The ability to test for water is important in the identification of unknown liquids. There are a couple of methods of identifying the presence of water, and the easiest is water-finding test paper (**FIGURE 10-24**). This paper changes from white to violet in the presence of pure water. When the paper changes to any color, there is water in the sample. The fact that the paper does not change colors does not mean water is absent, as some corrosives will bleach out the indicating dyes. Most corrosives have some amount of water in them, so if you get an extremely high or low pH, the water test may not be conclusive. A better method to check for water content is to break up some Alka-Seltzer™ tablets and put a pea-size amount in a test tube. Bubbling indicates the presence of water. To be assured that this test is successful, open the foil, and break up the tablet just before the test, to avoid its breakdown

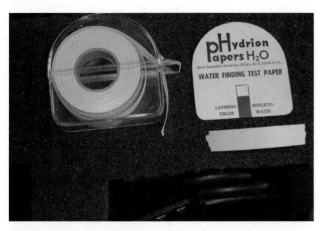

FIGURE 10-24 Water-finding test paper, which turns purple in the presence of water.
Courtesy of Christopher Hawley.

by exposure to humid air. Another method to check for water is water-finding paste, commonly used by gas station owners to check for water in their storage tanks. The paste is spread on the dipstick and indicates when water is present.

PCB Oil Tests

It is not a common issue, but it is one that responders may occasionally encounter, and that would be an issue with polychlorinated biphenyls, commonly referred to as PCBs. This was a material used as a coolant in transformers and electrical equipment, including fluorescent light ballasts. PCB oil was used in products that were produced prior to 1977. Its common names were Aroclor, Abestol, Askarel, and Clophen. These electrical items were supposed to be removed from service, but they may still be in existence. Schools are a likely location for fluorescent light ballasts that may contain PCBs. Newer ballasts and transformers will have a label that states they are PCB free. A gas chromatograph/mass spectrometer (GC/MS) device can be used to test an oil, but there is a colorimetric test that can be used. This test uses mercuric nitrate, which adds some disposal challenges. Dexsil Corporation makes these tests, and they offer three kits, based on the parts-per-million level. They offer kits that test 20, 50, or 500 ppm. Less than 50 ppm is considered to not be PCB contaminated. These tests identify chlorine, so any material with chlorine present will cause a positive reaction. These tests are also easily overwhelmed by more than 42 percent chlorine and will have false positives above that percentage. Water greater than 5 percent, alcohols, acetone, ketones, and acid may destroy the reagent. These kits have a shelf life of 1 year.

Infrared Thermometers

Infrared thermometers are useful in drum dumps to determine the temperature of the materials in the drum and to determine the liquid levels. Guided by a laser beam, the thermometers can determine the temperature of a surface from a considerable distance away. The exact distance is determined by the size of the target spot to be read. The target spot at 6 feet (1.8 m) is 2.4 inches (6 cm) with the Raytec S18 model. The temperature range that these thermometers can read is −25 to 1000°F (-31 to 538°C) (**FIGURE 10-25**).

Thermal Imaging Cameras (TIC)

These devices are typically used by firefighters to "see" in smoke-filled environments, but these devices are useful to check the liquid or solid levels within sealed and opaque containers (**FIGURE 10-26**). Propane tanks are supposed to only be filled to 80 percent of their capacity and present a risk if they are overfilled. One way to check if a propane tank is overfilled is to check its pressure, which requires that you come in contact with the container to check the pressure. You can use a TIC and keep your distance to check the level of the propane.

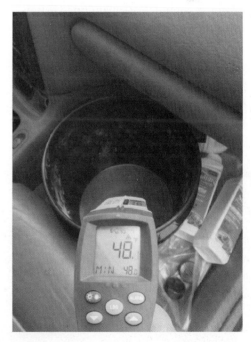

FIGURE 10-25 An infrared thermometer can determine the temperature of a material. In this instance, it is being used to determine the temperature of a chemical soup that was used in a chemical suicide.
Courtesy of Gary Sharp.

FIGURE 10-26 A thermal imaging camera (TIC), which can be used to determine the level of liquid in a container.
© Jones & Bartlett Learning. Photographed by Glenn E. Ellman.

FIGURE 10-27 A typical drum dump.
Courtesy of Maryland Department of the Environment Emergency Response Division.

Drum Sampling

HazCat and Spilfyter™ strips come in handy in drum sampling. Abandoned drums or what is commonly referred to as a drum dump, although not frequent, still occurs across the country (**FIGURE 10-27**). The dumping of drums is a serious environmental crime and should be handled as such. One of the issues with drum dumps is how they are handled (**FIGURE 10-28**). The chances of someone ensuring compatibility and proper packaging are slim whenever they dump the drums off the side of the road or wherever.

Responders should leave abandoned drums in place until they have been sampled and the contents fully

FIGURE 10-28 Pulling some samples from a drum dump using a drum thief.
John Emminzier Jr., Department of Energy & Environment District of Columbia.

Rock-n-roll drums, or drums that contain materials that are reacting, are an infrequent occurrence, but they do happen, so you should plan for them. The most important consideration is isolation of the event, as the drum could rocket a considerable distance if it were to rupture or explode. The second consideration is to determine what the contents are. In two events that Baltimore County's team handled, the mixture in the drums was ink and cleaning fluids, and the others were paint and paint waste. In one case, there was a single 55-gallon (208-liter) drum rocking and rolling on a loading dock. The facility backed up to a neighborhood, so letting the event run its course was not an option. We consulted with our chemist, who advised a cocktail mixture of aluminum powder and alkaline waste to stop the reaction. Two hazardous materials team members approached the drum, and while one held the drum, the other opened it, slowly venting the pressure. Once the drum was opened, the cocktail mixture was added, and the reaction stopped. In the other incident, two drums out of 30 were rocking and rolling, so our chemist determined a different cocktail mixture, and all the drums were treated. If remote opening is possible, that is always recommended, but speed is of the essence in this high-risk mitigation, and having immediate access to a chemist is essential.

characterized. A smart dumper uses a transmission fluid drum, labeled as such, and fills the drum with waste material, which to make the point we will say is a heavy metal contaminated dilute sulfuric acid. On the top of the sulfuric acid the dumper places a quart or so of transmission fluid, which floats on top of the acid. The dumper then finds a suitable location to dump the drum. The more time he or she has to get away, the less likely he or she is to get caught. If the dumper chose a library in the downtown area to put this drum, eventually someone would phone the authorities about the drum, and someone would investigate the drum. Looking at the drum, he or she would see the transmission fluid label or other nonhazardous markings, and if he or she opened the drum, he or she would only see the transmission fluid floating on top. Most responders would be inclined to call this a drum of transmission fluid and leave the drum for later pickup, but all the while the sulfuric acid is eating away at the steel drum and generating hydrogen gas. At some point, the drum is going to fail, either due to increased pressure or simple corrosion, releasing the metal-contaminated sulfuric acid. One important issue is that most responders would assume it was transmission fluid in the drum; thus, it is unlikely it would be isolated or picked up quickly.

It would be a low priority, even though the whole time that drum sits there its potential for a catastrophic release is increasing. Sampling the drum with a drum thief, which is a drum-sized straw, would take about a minute and would provide valuable information. Taking a sample and characterizing, in addition to using HazCat, would quickly indicate a mixture and that the bottom layer was a strong acid. At a minimum, the drum would have to be transferred or overpacked into an appropriate container and the hazard controlled.

Opening drums presents a risk, but most drums that are going to be extremely risky to deal with provide several clues that there is a problem. Drums that are swelled, bowed up, expanded, or obviously under pressure present additional risk, and other methods of opening, such as remote opening, should be considered.

SAFETY TIP

Responders should leave abandoned drums in place until they have been sampled and the contents fully characterized.

Most drums that have been sitting outside have some pressure. Unscrewing the bung cap very slowly, a quarter inch at a time, will safely vent the pressure. While doing so, the air monitors should be close to the bung to evaluate the released gas. There are several ways to open reacting drums, but location, exposures, and timing are major factors in determining the method

used. Opening by humans is extremely risky and should be avoided. However, when citizens' lives are at risk and time is of the essence, a minimum of responders must be used. The local bomb squad should have a robot, which can be used for this purpose. The robot has movable arms and can open some containers; it can also puncture containers or shoot them with a shotgun slug or high-velocity water.

SAFETY TIP

Drums that are swelled, bowed up, expanded, or obviously under pressure present additional risk, and other methods of opening, such as remote opening, should be considered.

The ability of the bomb technician to open drums varies, as experience levels vary from one bomb technician to another. Some of them can untie shoes using a robot, while others have some difficulty handling fine tasks. Robots require setup time, in addition to response time, which is a major concern with a reacting drum. Commercially available remote drum opening devices allow a drum to be punctured while capturing any resulting vapor or liquid release. However, these remote devices must be placed on top of the drum by humans. Their use requires practice, and it is not likely that a bomb robot would be able to place such a device on top of a drum. Having immediate access to a chemist who can help develop the cocktail required to stop a reaction is essential, preferably one who is familiar with hazardous materials operations.

Most hazardous materials teams use poly drum thieves for their sampling, as glass thieves tend to break. If you are sampling for evidence, you must use glass, and it is only used once. Most teams that are collecting evidence use **coliwasa** tubes, as they can give an exact cross-sectional representation of what is in that drum. There are both poly and glass coliwasa tubes. Anything used for evidence must be certified clean, kept sealed, then used one time to collect a sample. Once the sample is collected, it can be tested the same as any other unidentified liquid.

After-Action REVIEW

IN SUMMARY

Detection devices such as mercury detectors, HazCat kits, and test strips are a valuable resource for identifying unknown materials. Both low-cost and expensive technical devices are available to responders. Some initial detection can be done with the low-cost devices, and

then further characterization can be done outside the hot zone with the more technical kits. Developing a system that combines many different technologies can be difficult, and Chapter 12 can assist in the development of a sampling system.

KEY TERMS

Chameleon A colorimetric test kit for a variety of chemicals.

Chemical classifier A testing strip that includes five tests; a companion test strip is the wastewater strip.

Coliwasa A tube used to collect samples from a drum or tank; formal name is composite liquid waste sampler.

Fourier transform infrared device Detection device that uses infrared light to detect solids, liquids, and gases.

HazCat™ kit A chemical identification kit that can be used for solids, liquids, and gases.

Raman spectroscopy Detection technology in which a laser is used to excite the sample.

Wastewater strip A strip that does some additional tests beyond the chemical classifier.

Review Questions

1. What is the technology behind FTIR?

2. What is the technology behind Raman?

3. What should be tested with care when using a Raman-based device?

4. What is dangerous about Raman?

5. What can FTIR not detect?

6. When using a HazCat kit, how much sample should you test?

7. Mercury presents what type of exposure issue?

CHAPTER 11

Terrorism Agent Detection

OUTLINE

LEARNING OBJECTIVES

Upon completion of this chapter, you should be able to:

- Identify common terrorism threat agents
- Describe common detection methods for terrorism agents
- Describe sampling strategies for terrorism agents
- Identify methods to collect samples following evidentiary procedures
- Describe the technology that drives chemical and biological detection

Hazardous Materials Alarm

Hazardous Materials Alarm Explosives in the Basement—The police department was called by a car wash that was under explosive attack. When the police officer arrived at the car wash, he too was taking explosive rounds bursting in the air over his head. He determined the trajectory of the devices and went to that location. When he went into the backyard, he noticed a 50ish man with a smoking polyvinyl chloride (PVC) pipe in his hand. When asked by the officer where he got the fireworks, the obviously intoxicated man—not recognizing that the person he was talking to was a police officer—replied "Come on, I will show you." When they went into the basement of his house, there were explosives, bomb-making supplies, and many ashtrays and empty beer cans all over the place. The police officer arrested the gentleman and retreated out of the basement, calling for the bomb squad. When it arrived, the bomb squad needed help identifying some of the chemicals in the basement. Although there were many things to be concerned with throughout the basement, some very serious explosives included pentaerythritol tetranitrate (PETN) in the freezer. There were several corrosives, including concentrated nitric acid. We tested the pH, and it was about 0, and then using a Spilfyter™ chemical classifier (described in Chapter 10), we identified that it was also an oxidizer, exhibiting the characteristics of nitric acid. The bomb technicians gathered up the explosives, and we gathered up the many chemicals for disposal.

1. What is the weapon of choice that terrorists are likely to use?

2. What are the basic detection devices that should always be used when entering a residence that is suspected of housing a terrorist?

3. Who should be consulting when you may have to handle evidence?

Introduction

With the increasing concern over terrorism, responders need to have the ability to detect chemical and biological warfare agents, some of which are shown in **TABLE 11-1**. Radiological materials, another potential terrorist weapon, are discussed in Chapter 9. Of the variety of detection devices for chemical and biological warfare agents, some are simple to use, and others are very complicated

TABLE 11-1 Chemical Warfare Agents		
Agent Class	**Common Agents**	**Military Designation***
Nerve	Sarin	GB
	Tabun	GA
	VX agent S-[2-(Diisopropylamino)ethyl] O-ethyl methylphosphonothioate	VX
	VE agent S-(Diethylamino)ethyl O-ethyl methylphosphonothioate	VE
	Soman	GD
	Cyclosarin	GF
	Ethyl sarin	GE
	Amiton or Tetram	VG
	VM agent S-[2-(Diethylamino)ethyl] O-ethyl methylphosphonothioate	VM

TABLE 11-1 Chemical Warfare Agents (*continued*)

Agent Class	Common Agents	Military Designation*
Blister (vesicants)	Sulfur mustard	H
	Distilled sulfur mustard (Bis(2-chloroethyl) sulfide)	HD
	Bis (2-chloroethyl)ethylamine	H1
	Bis (2-chloroethyl)methylamine	H2
	Tri (2-chloroethyl)amine	H3
	Lewisite (2-chlorovinyldichloroarsine)	L
	Mustard lewisite	HL
	Phosgene oxime	CX
	Ethyldichloroarsine	ED
	Methyldichloroarsine	MD
	Phenyldichloroarsine	PD
Riot control (irritants)	Tear gas	CS
	Mace (chloroacetophenone)	CN
	Capsaicin (pepper)	OC
	Bromoacetone	BA
	Bromobenzyl cyanide	CA
	Dibenzoxazepine	CR
Vomiting agents	Adamsite	DM
	Diphenylchloroarsine	DA
	Diphenylcyanoarsine	DC
Incapacitating	3-Quinuclidinyl	BZ
	Phencyclidine	SN
	Lysergic acid diethylamide	LSD
Choking	Chlorine	CL
	Chloropicrin	PS
	Diphosgene	DP
	Phosgene	CG
Blood agents	Hydrogen cyanide	AC
	Cyanogen chloride	CK
	Arsine	SA

*Many detection devices still use the abbreviation for the military designation for the material, so it is important to have a quick reference guide to assist.

© Jones & Bartlett Learning

(**FIGURE 11-1**). One of the difficulties with the use of these devices is maintaining user competency and proficiency; responders need to train and use these devices frequently.

Although these problems exist with other devices, there are two major issues with chemical and biological warfare agent detection devices. Two terms, "**false positive**" and "**false negative**," are used when discussing warfare agent detection devices. False positives occur when a detection device reports the presence of a material and none is present. A false negative occurs when the device reports that a material is not present when it is. False positives require responders to take a more protected posture than is necessary.

FIGURE 11-1 The GC/MS requires training and some chemistry background to assist in the interpretation of the results.

Courtesy of Christopher Hawley.

False negatives, on the other hand, may cause response teams to be less self-protective, creating a situation in which they may become exposed to a material. False positives, although less desired, are much better than false negatives. Most chemical detection devices provide false positives, and none are 100 percent accurate. Biological detection devices are more prone to false negatives, which are discussed later in the chapter.

Methods of Detection

There are five basic methods of detection of warfare agents. The basic methods are:

- Colorimetric detection/test strips
- Colorimetric sampling
- Direct-read instruments
- Biological testing
- Field/lab analysis

SAMPLING AND EVIDENCE COLLECTION

One of the major needs of local law enforcement agencies and the FBI for the prosecution of a terrorism case is purity of evidence. For evidence to be used successfully in a prosecution, it must be pure beyond all doubt. If there is any suspicion that it has been tampered with, altered, or contaminated, it can adversely affect the outcome of the prosecution. When using detection devices to check for the presence of weapons of mass destruction (WMD) materials, there is great potential for evidence to be destroyed. The best way to proceed is to coordinate efforts with the local FBI WMD coordinator, who can assist with the preservation of evidence.

Several questions arise regarding evidence. When are responders dealing with a potential crime? When should they concern themselves with evidence? At what point do they follow evidentiary procedures? The easiest answer is to follow the rules for every situation. If a hazardous materials response team has been called into a situation, regardless of circumstances, there is a good likelihood that lawyers and the courts may get involved. The case may not be a criminal one but rather a workers' compensation lawsuit or other civil matter. Responders may be called to testify about their activities or about the results of their testing. The responders or their agency may also be the target of a lawsuit, which requires a defense.

The process of evidence collection is very process driven, but procedures can be implemented to preserve potential evidence. This is a brief overview of the process and some items to consider.

First, the gold standard for any courtroom is laboratory analysis, so evidence must be available for analysis by a laboratory. Only the laboratory can positively identify a material. Even with all the sophisticated devices described by this text, field analysis is considered to be a presumptive identification. Federal courts only recognize

laboratory results. Some state and local court systems may accept the results of characterization, but there must be a documented collection process in place. Not every incident requires material be sent off to the laboratory, but steps should be taken to preserve evidence in the event that laboratory testing is necessary. To accomplish this task, take at least half of the available material, and place it into an appropriate container. If there is only a small amount of material, then seriously consider sending all of it to the laboratory. The laboratory needs only small quantities of materials; in most cases, 0.7 ounces (20 mL) of a liquid is more than sufficient. For biological materials, a couple of swabs that have come in contact with the material and been properly contained may suffice, or you can collect a gram or two of material. With the remaining portion of the evidence, you can perform any kind of street test; just keep the evidence portion pure. It is best to check with your local or state crime laboratory for their specific criteria.

The collection tools and containers all must be certified clean and/or sterile, and they should have lot numbers (**FIGURE 11-2**). They must come from sealed boxes, which have been tracked and have documentation about their certification. Each new sampling point or test requires a new tool and container. The process must be documented: Who performed which sampling, with dates, times, and specific notes on the actions taken. Each evidence or sampling container must be numbered, and the collector's initials, the date, and the time must be included.

The process of collecting evidence is known as sampler–assistant sampler. The assistant sampler handles only clean and pure items and equipment, mainly providing the tools to the sampler. The assistant sampler should never come in direct contact with the sampler or any

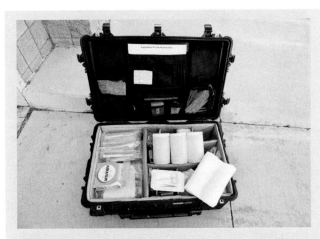

FIGURE 11-2 This is an example evidence kit, which is modeled after the FBI's hazardous evidence collection kit.
© Jones & Bartlett Learning. Photographed by Glenn E. Ellman.

evidence. The sampler is the one who samples the material being collected for evidence. This person:

- May touch potentially contaminated surfaces and containers when opening containers to sample.
- Performs the actual evidence collection, using tools to pick up the evidence.
- Takes great care not to cross-contaminate or come in direct contact with potential evidence and a contaminated tool.

Both the sampler and the assistant sampler wear at least two pairs of gloves and bring a quantity of new gloves with them because they change them frequently (**FIGURE 11-3**). The sampler establishes a clean work area and lays out the equipment. Each piece of evidence is screened for fire, corrosive, toxic, oxidizer (explosive), and radioactive hazards. When the containers are opened, the process is repeated. In this case, one pipette is used to collect the material from container 1. The sampler pipettes liquid into a sample container, places the lid on the container, and puts it into an overpack container. If there is enough material left in the pipette, or they can pipette more, they place a small amount onto pH paper to check for corrosivity. If any other tests are to be to run, they use the pipette to collect the sample. Dipping pH paper into the liquid

FIGURE 11-3 The sampler is placing a small amount of a material into a collection container.
Courtesy of Christopher Hawley.

in the container contaminates the evidence, rendering it useless. After being used for testing, the pipette is thrown away, as are the assistant sampler's outer gloves. Once they don a new set of gloves, they use a new pipette and sample jar from the assistant sampler. They collect liquid from container 2 following the same procedure. They change their gloves and don a new pair. Using a scoop or other certified clean tool, they scoop up some of the powder and place it in a clean container. A small amount of the powder is also placed in another sample jar, and a small amount of water is added. They test the pH of the sample in solution and note any reactions or bubbling. The tool is disposed of, and the gloves are changed. All the evidence is marked, and written documentation of the specific activities is taken.

When a large amount of potential evidence is needed, the process can be overwhelming, but following it helps eliminate any errors and preserves evidence. Responders face a great potential to become involved in a court case, and following the proper procedures every time minimizes their risk in the courtroom. The one time they do not follow evidentiary procedures will be the time they are grilled in the courtroom for contaminating the evidence. The FBI and local law enforcement agencies have response teams that collect evidence in these situations, and they can provide great assistance when responders are trying to process a potential crime scene.

These methods mirror those for detecting dangerous substances in the civilian world. The most common direct-read instruments are the ChemPro and the LCD 3, which provide readings much like the photoionization detector, although the technology to ionize the gas is different (**FIGURES 11-4** and **11-5**).

Colorimetric Detection/Test Strips

Every hazardous materials team in the country should have a minimum method of detecting warfare agents.

One of the simplest methods for detecting chemical agents is the use of a combination of test strips.

M8 paper is like pH test strips for warfare agents, and it indicates for nerve and blister agents. It provides a separate test for nerve agent VX, and through a color change indicates which type of agent is present. The M8 test kit comes with 25 test sheets, and a color-indicating chart is provided in the front of the test booklet (**FIGURE 11-6**). A liquid sample is needed for the test; it requires at least 0.02 mL (0.0007 oz) for the test to work and has a 30-second reaction time.

M9 paper is better known as M9 tape, as it is a roll of tape with an indicating layer on the outside surface (**FIGURE 11-7**). It is commonly placed on vehicles,

FIGURE 11-4 The ChemPro 100i detection device, which can detect chemical warfare agents and some toxic industrial materials using IMS and electrochemical sensors.
© Jones & Bartlett Learning. Photographed by Glenn E. Ellman.

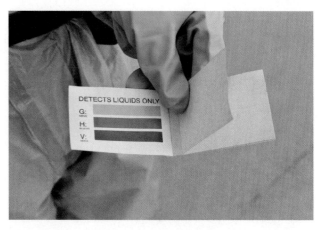

FIGURE 11-6 M8 paper used to detect nerve and blister agents.
© Jones & Bartlett Learning. Photographed by Glenn E. Ellman.

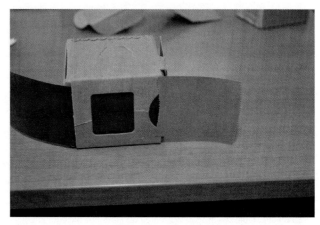

FIGURE 11-7 M9 test paper, which is in tape form, indicates red in the presence of liquid nerve and blister agents.
Courtesy of Christopher Hawley.

FIGURE 11-5 The LCD (JCAD) device, which detects chemical agents.
Courtesy of Christopher Hawley.

boots, gloves, and other parts of the responder. M9 indicates the presence of nerve and blister agents through a red color change. M9 does not differentiate between nerve or blister; it only indicates the presence of a warfare chemical. M8 and M9 are used together as they react to differing sets of false positives. M9, like M8, has a lot of false positives. M9 is useful because it is more sensitive than M8 paper and will indicate first. In low levels, there may not be a distinctive color change, but there may be pinpoint color changes on the tape. The pale green paper is impregnated with indicating dyes that turn pink, red, reddish brown, or red-purple after being exposed to liquid.

The **M256A1 kit** and the **C2** kit are military devices that can be used to detect warfare agents. The M256A1 kit is used by the U.S. military, and the C2 kit is used by the Canadian Defense Forces (**FIGURE 11-8**). Both kits have cumbersome instructions and are difficult to use, especially in personal protective equipment (PPE). They require regular training, and they employ wet chemistry as the detection method. The M256A1 kit detects nerve, blister, and blood agents in the air, as opposed to M8 and M9, which detect only liquids. The M256A1 kit has a low cost but is difficult to operate. It takes about 20–25 minutes to operate and a great many steps. The user is required to follow the steps in order and the timing for each. The test card must be held in a variety of configurations for specified amounts of time. Chemical-filled vials are broken open at various times during the test. The instructions are printed in very small type and are difficult to read. It is recommended to make an enlarged photocopy of them and laminate them for ease of use. A stopwatch is recommended to

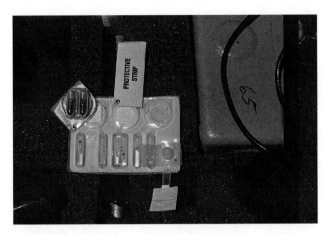

FIGURE 11-8 The M256A1 test kit indicates nerve, blister, and blood agents in the vapor form.
Courtesy of Christopher Hawley.

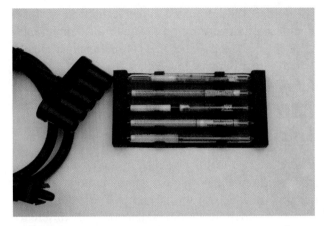

FIGURE 11-9 These tubes will detect a range of warfare agents at one time.
Courtesy of Christopher Hawley.

make the timing easier because several steps must be timed. However, this is the only nonelectric detection device that can detect vapors in the air.

Colorimetric Sampling

One of the best methods for detecting warfare agents without using an electronic device is to use colorimetric sampling tubes (**FIGURE 11-9**). Only one company, Dräger, makes colorimetric tubes for specific warfare agents, but most companies provide tubes for the hydrolysis products of the agents. There are two primary approaches in use for colorimetric tubes: Individual and simultaneous analyses. The individual chemical analysis involves selecting specific tubes for each threat and analyzing independently. The simultaneous analysis tube kits sample for a variety of agents at one time. Obviously, with time being of the essence, the multisampling system has the advantage. The individual tests allow for more sampling flexibility and assist in the final identification, so a combination strategy should be employed. The multitest provides one shot at an identification, while the individual tubes come in boxes of 10 tubes. One of the biggest disadvantages of the **Civil Defense Simultest** (CDS) system is time. The system requires 50 strokes with the bellows pump. Each pump stroke takes up to 1 minute, so the test takes more than 50 minutes to complete. Dräger makes an electronic pump system that automates the pump strokes, as opposed to responders performing the pump strokes. The pump does not perform the tests any faster, but it eliminates the need to perform the pump strokes by hand. It is very important when using colorimetrics to read the instructions, but it is even more important with the warfare agent tubes, as in some cases they are used differently from regular

colorimetric tubes. In most cases, the tubes used to detect the warfare agents are detecting the hydrolysis products of the warfare agent and not the agent itself, so a positive does not mean that the warfare agent is present, only the fact that a hydrolysis agent is present. One major plus for the colorimetric tubes is that they have the fewest false positives when dealing with the warfare agents, although there are some cross sensitivities. Generally, if they indicate the presence of warfare agents, they are most likely present. A listing of the Dräger colorimetric tubes for warfare agents is provided in **TABLE 11-2**.

TABLE 11-2 Warfare Agents and Colorimetric Tubes	
Warfare Agent	**Dräger Colorimetric Tube**
Lewisite	Organic arsenic compounds
S-mustard	Thioether
N-mustard	Organic basic nitrogen compounds
Sarin	Phosphoric acid esters
Soman	Phosphoric acid esters
Tabun	Phosphoric acid esters
Cyanogen chloride	Cyanogen chloride
Phosgene	Phosgene
Hydrogen cyanide	Hydrocyanic acid

More recently, automated systems that can perform simultaneous identification and quantification for multiple chemistries have become commercially available, which were discussed in Chapter 10.

Direct-Read Instruments

There are two major divisions of direct-read instruments: Handheld, portable instruments and those that are prepositioned for special events. The technology of these two divisions is comparable; the only difference is the box the detector is in.

LISTEN UP!

One major plus for the colorimetric tubes is that they have the fewest false positives when dealing with the warfare agents. Generally, if they indicate the presence of warfare agents, they are most likely present.

Prepositioned equipment is usually radio or hardwired controlled to a central location. Direct-read instruments and their method of operation are listed in **TABLE 11-3**.

Chemical Agent Detection Devices

Several types of chemical agent detection devices use a variety of detection technologies, which include **ion mobility spectrometry (IMS)**, flame spectroscopy, and combinations of these types. One technology that was on the market was surface acoustical wave (SAW), but there are no current models using this technology. Several emerging technologies are discussed later in this chapter. Regardless of the technology, a response team should not rely on only one. A combination is highly recommended. Each of the technologies has its problems, but, when used in combination, they help eliminate false positives, which relate to a device's accuracy.

Except for flame spectroscopy, the development of individual chemical fingerprints in the library, sometimes referred to as the spectra or plasmograms, is the most difficult aspect of manufacturing the devices (**FIGURE 11-10**). Because the manufacturer develops the fingerprints for a variety of chemicals and the software to place them in the device, what the device reports as a positive indication, or "hit," varies according to the manufacturer. When a sample material has the characteristics, or fingerprint, of a chemical warfare

TABLE 11-3 Direct-Read Instruments

Device	Agent Detected	Mode of Operation
LCD	Nerve, blood, blister, choking, and select toxic industrial chemicals	Emergency response
Sabre 5000	Nerve, blister, explosives, drugs, and select toxic industrial chemicals	Emergency response
LCD-Nexus	Nerve, blood, blister, choking, and select toxic industrial chemicals	Prepositioned, vehicle mounted and/ or emergency response
ChemPro 100i		
GID-3	Nerve, blister, blood, and choking	Prepositioned, vehicle mounted and/ or emergency response

© Jones & Bartlett Learning

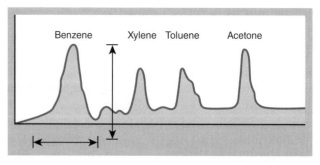

FIGURE 11-10 Ion mobility fingerprint provides peaks, which are measured for height and width and compared to a library of other peaks. If the peaks match or are close, then the meter alarms.
© Jones & Bartlett Learning

agent, the device is likely to report the presence of a warfare agent. The identification of chemical warfare agents is very difficult because many have very low vapor pressures and do not produce the large quantities of vapor that the electronic detection devices require. Chemical warfare agents are also very toxic at very low levels, and symptoms can occur in the parts-per-billion range. Manufacturers face the difficulties of making a device that can accurately detect very low levels of vapor from a very toxic liquid that does not produce large quantities of vapor. Although the manufacturer has a desire for a high level of accuracy, they also do not want to miss any potential toxic materials. The fingerprints are designed for U.S. military chemical agents, which are of a high quality. Someone who cooks a chemical warfare agent in his or her home may not meet the same quality standards. The detection device should have the ability to detect a chemical material that has some of the same characteristics as the pure material. It is a difficult situation for the manufacturer to handle. The analysis of chemical materials allows for close matches to a pure fingerprint to be reported as a match. How close a match determines how accurate the device is.

Each of the three major technologies has advantages and disadvantages, but the use of at least two of them provides an accurate system. The IMS instruments provide the ability to detect low levels of materials but tend to return false positives. SAW units had provided more accuracy, but they couldn't detect materials at low levels. The flame spectroscopy units also can detect at low levels, but they are not specific to a material. Instead, they only address the elemental components within the material. Each of these technologies reacts to a differing set of false positives. If one unit reports a positive and one reports a negative, then the results are likely negative. If both report a positive, the results are likely positive. If both report a negative, the results are likely negative, or the amount of material present is lower than the detection threshold.

LISTEN UP!

One of the best methods for detecting warfare agents without using an electronic device is to use colorimetric sampling tubes.

Principles of Ion Mobility Spectrometry

Ion mobility spectrometry is one of the most common detection technologies for chemical warfare agents. IMS has a great ability to detect chemical materials, specifically those that are ionizable. Most older IMS units use a radiation source, typically nickel-63 or americium, as the source of energy for ionization. The Smiths LCD 3.3 (JCAD) device and the Bruker µRAIDplus use corona discharge, which is further explained in Chapter 7.

Although responders may have a concern about the radiation source being present in the detection device, the device is safe to use. The americium radiation source in household smoke detectors is the same as the one used in the Environics ChemPro 100i device. Some Smiths devices and Bruker devices use a nickel-63 source. Some of these devices may need to have periodic wipe testing to ensure the source remains sealed. Over the last few years, manufacturers have started using nonradioactive ionization sources to minimize the regulatory burden from the storage, use, and transport of radiological materials.

How the IMS device works is comparable to other ionization units. The sample gas or sample material is brought into the sensing chamber by an internal pump (**FIGURE 11-11**). One advantage of IMS is that it can identify solid materials in addition to gases and vapors. To detect solid materials, a small section of detection paper is used to pick up the sample material, which is then placed into the device. The device heats the sample, forcing the resulting vapors into the device. In the sampling of a gas or a vapor, the heater is not used, and the material enters the sample chamber. Once in the sample chamber, the nickel-63 bombards the sample with beta emissions, ionizing it. Once ionized, the various sample components move to a sample gate, which focuses the sample. The sample gate opens and closes depending on the charge to the sample. The ionized sample moves past the gate to the drift tube, which provides the mobility aspect of ion mobility spectrometry. The sample moves to a sensor at the other end of the drift tube, which has an electrical field and determines the length of time the sample took to move through the tube. Measurements of the electrical activity and the drift time produce a peak, which can be referred to as a fingerprint, because it is specific to a given molecule. The peak developed is compared to peaks in the library, and the device looks for a match. If a match is made, the device reports that it found a particular material. The detection of the lethal nerve agent VX is the most problematic and usually presents most of the false positives. This is due to the low vapor pressure of VX and the inherent desire of the manufacturer to make sure the device finds this super toxic material. When the manufacturer designs algorithms, it has two choices: Tight or loose. A tight algorithm is one that reports a positive when the fingerprints are very closely matched. A loose algorithm is one that provides a positive for chemicals that are fairly close or have some of the characteristics of the original algorithm.

The **SABRE 5000** is a combination unit that can detect a wide variety of materials. It can detect chemical warfare agents, such as nerve and blister agents. Other devices can detect explosives and drugs in addition to chemical warfare agents. Some devices use other technologies to detect groupings of industrial chemicals, such as

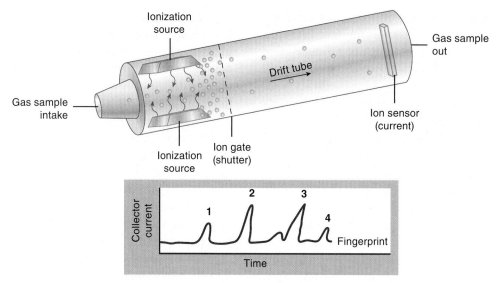

FIGURE 11-11 Schematic of an ion mobility drift tube. The sample gas enters the device and is ionized, typically with Ni-63 beta radiation. The ion gate allows the charged ions to pass into the drift tube. The travel time down the tube is recorded at the ion sensor (collector), which records the fingerprint. Peaks 1–4 all represent the chemical makeup of the sample gas. Peak 1 is a reference peak.

© Jones & Bartlett Learning

chlorine, ammonia, and hydrogen cyanide. To detect industrial chemicals, most of the time these devices use electrochemical sensors (the same as those discussed in Chapter 5), or they use IMS. Most manufacturers are forced to detect only a small number of materials using IMS. Although the technology can detect a wide variety of substances and small amounts of chemicals, manufacturers have not designed the algorithms for other chemicals. The advantage to IMS is its ability to detect low levels, down to parts-per-billion levels. Its basic disadvantage is that it is slow to respond to low levels, some units taking up to 1½ minutes to respond to low levels of chemical warfare agents.

IMS Detection Devices

There are several IMS devices on the market because they are the most common type of detection devices. The LCD 3.3 is the first responder version of the military Joint Chemical Agent Detection (JCAD) device, manufactured by Smiths Detection (**FIGURES 11-12, 11-13**, and **11-14**). Both devices use corona discharge as their method if ionization. The ChemPro 100i is an IMS device. Environics of Finland makes the **ChemPro100i**, which detects chemical warfare agents and toxic industrial materials. The ChemPro 100i has a group of detection sensors inside the device, transparent to the user, that is known as an orthogonal detection device. This means it uses a variety of sensor types that

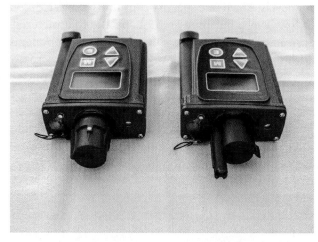

FIGURE 11-12 The JCAD version of the LCD, which is an IMS device that uses corona discharge to detect chemical warfare agents.

Courtesy of Christopher Hawley.

feed into an electronic "fusion center" where it makes decisions about the gas sample. The heart of the device is an aspirated IMS sensor, but it also contains semiconductor sensors (SCS) and metal oxide sensors (MOS). The ChemPro has several libraries for the detection of unidentified materials. One of the best aspects of the ChemPro is its trend analysis function in which the device functions much like a photoionization detector (PID). When in this mode, the screen shows a line, such as an EKG monitor. When there are no threat materials in the air, the line stays flat (**FIGURE 11-15**).

FIGURE 11-13 The LCD has a nozzle that allows pinpoint sampling.
Courtesy of Christopher Hawley.

FIGURE 11-14 The LCD (JCAD) has a sieve pack (white) that contains the dopant to calibrate the device. The black pack is a carbon pack, which should be placed in the device when it is not in use. The sieve pack requires changing after a certain number of hours.
Courtesy of Christopher Hawley.

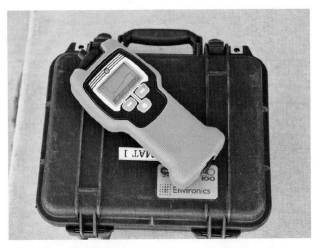

FIGURE 11-15 One of the best features of the ChemPro is its trend analysis line, which alerts the user to contaminants in the air.
© Jones & Bartlett Learning. Photographed by Glenn E. Ellman.

When a threat is detected, the line trends upward, indicating the presence of something in the air. If the material is in the library, it will alarm and indicate the material present. The device has 12 libraries, which are provided in **TABLE 11-4**. This device has incredible capabilities for the detection of WMD materials, meth lab threats, fire smoke threats, irritants, and a range of common industrial materials. Responding to a mall or a school for some type of chemical release was always frustrating, as experienced responders knew the release was mace or pepper spray, but colorimetric tubes were not helpful unless they arrived quickly. This detector has solved a lot of hazardous materials issues when dealing with this kind of event. Although sold and marketed as a WMD device, the trend analysis function and the fire smoke and irritants libraries are well worth the money.

Bruker Daltonics makes several WMD detection devices, but the one most used for civilian work is the **RAID-M-100**, which detects chemical warfare agents and toxic industrial materials. Their devices are manufactured in Germany and are imported to the United States.

The SABRE 5000 is designed for general first responder use and can detect chemical warfare agents and some toxic industrial chemicals. The toxic industrial chemicals include phosgene, nitric acid, ammonia, chlorine, hydrogen cyanide, sulfur dioxide, ethylene oxide, hydrogen chloride, and hydrogen fluoride. The SABRE 5000 (or any comparable device) is one of the best WMD detection devices that a hazardous materials team should own. If you were looking to identify potentially hazardous materials that could be used in a WMD scenario, the SABRE 5000 can detect most

TABLE 11-4 ChemPro 100i Libraries

TIC Auto Classifier (First In)	
Screen Indication Icon (Group)	**Detects**
Ammonia	Ammonia
Chlorine	Chlorine
Hydrogen cyanide	Hydrogen cyanide
Hydrogen sulfide	Hydrogen sulfide
Oxidizers	Fluorine, nitrogen dioxide
Hydrides	Arsine, diborane, phosphine
Acids	Hydrogen bromide, hydrogen chloride, hydrogen fluoride, nitric acid, phosphorus trichloride
Organic	Acrylonitrile, allyl alcohol, carbon disulfide, ethylene oxide, formaldehyde, methanol
Toxic industrial chemicals	Boron, trichloride, boron trifluoride, cyanogen chloride, phosphorus oxychloride, sulfur dioxide
Chemical	When it senses something in the air and cannot identify it
First Responder Mode	
Screen Indication Icon (Group)	**Detects**
Toxic	Acrylonitrile, allyl alcohol, ammonia, arsine, benzene, boron, trichloride, boron trifluoride, carbon disulfide, carbon monoxide, chlorine, cyanogen chloride, diborane, ethylene oxide, fluorine, formaldehyde, hydrogen bromide, hydrogen chloride, hydrogen cyanide, hydrogen fluoride, hydrogen sulfide, methanol, nitric acid, nitrogen dioxide, phosphine, phosphorus oxychloride, phosphorus trichloride, sulfur dioxide
Chemical detected	When it senses something in the air and cannot identify it
TIC Classifier – Hydrides, Oxidizers, and Acids	
Screen Indication Icon (Group)	**Detects**
Oxidizer	Chlorine, nitrogen dioxide
Hydride	Ammonia, arsine, diborane, hydrogen sulfide, phosphine
Acids	Hydrogen bromide, hydrogen chloride, hydrogen fluoride, nitric acid, phosphorus trichloride
Organic	Acrylonitrile, allyl alcohol, carbon disulfide, ethylene oxide, formaldehyde, methanol
Toxic industrial chemicals	Carbon monoxide, cyanogen chloride, hydrogen cyanide, sulfur dioxide

TABLE 11-4 ChemPro 100i Libraries (*continued*)

TIC Confirm

Screen Indication Icon (Group)	Detects
Ammonia	Ammonia
Chlorine	Chlorine
Cyanogen chloride	Cyanogen chloride
Hydrogen cyanide	Hydrogen cyanide
Hydrogen fluoride	Hydrogen fluoride
Hydrogen sulfide	Hydrogen sulfide
Sulfur dioxide	Sulfur dioxide
Chemical detected	When it senses something in the air and cannot identify it

Chemical Warfare Agents—Sensitive

Screen Indication Icon (Group)	Detects
Nerve	Sarin, tabun, soman, cyclosarin, VX, organophosphate pesticides
Blister	Sulfur mustard, lewisite, nitrogen mustard
Blood	Hydrogen cyanide, cyanogen chloride
Toxic industrial chemicals	Chlorine, ammonia
Chemical detected	When it senses something in the air and cannot identify it

Irritants

Screen Indication Icon (Group)	Detects
Irritants	Mace, pepper spray, tear gas, CN, CS
Chemical detected	When it senses something in the air and cannot identify it

Clan Lab—Methamphetamine

Screen Indication Icon (Group)	Detects
Mask up (threat present)	Acetic acid, acetone, ammonia, benzene, ethers, ethyl acetate, ethyl alcohol, gasoline, hydrogen chloride, iodine, isopropyl alcohol, methyl amine, phosphine, phosphorus trichloride, Stoddard solvent, toluene
Chemical detected	When it senses something in the air and cannot identify it

(*continued*)

TABLE 11-4 ChemPro 100i Libraries (*continued*)	
Fire Smoke—Overhaul	
Screen Indication Icon (Group)	**Detects**
Mask up (threat present)	Acetaldehyde, acrolein, acrylonitrile, ammonia, benzene, carbon monoxide, formaldehyde, formic acid, glutaraldehyde, hydrogen bromide, hydrogen chloride, hydrogen cyanide, hydrogen fluoride, isocyanates (TDI, MDI), naphthalene, nitrogen oxides (NO, NO_2), sulfur dioxide, toluene, vinyl chloride
Chemical detected	When it senses something in the air and cannot identify it
CWA Standard	
Screen Indication Icon (Group)	**Detects**
Nerve	Sarin, tabun, soman, cyclosarin, VX, organophosphate pesticides
Blister	Sulfur mustard, lewisite, nitrogen mustard
Blood	Hydrogen cyanide, cyanogen chloride
Toxic industrial chemicals	Chlorine, ammonia
Chemical detected	When it senses something in the air and cannot identify it
CWA High Sensitivity	
Screen Indication Icon (Group)	**Detects**
Nerve	Sarin, tabun, soman, cyclosarin, VX, organophosphate pesticides
Blister	Sulfur mustard, lewisite, nitrogen mustard
Blood	Hydrogen ayanide, cyanogen chloride
Toxic industrial chemicals	Chlorine, ammonia
Chemical detected	When it senses something in the air and cannot identify it
CWA Identification for Pure Materials	
Screen Indication Icon (Group)	**Detects**
Sarin or tabun	Sarin or tabun
Soman or cyclosarin	Soman or cyclosarin
VX	VX

TABLE 11-4 ChemPro 100i Libraries (*continued*)

Screen Indication Icon (Group)	Detects
Sulfur mustard or lewisite	Sulfur mustard or lewisite
Hydrogen cyanide	Hydrogen cyanide
Chemical detected	When it senses something in the air and cannot identify it
Precursor Materials	
Screen Indication Icon (Group)	Detects
Nerve	Dimethyl methyl phosphonate, dimethyl phosphite, methyl phosphoric dichloride
Vesicant	Thiodiglycol
Chemical detected	When it senses something in the air and cannot identify it

© Jones & Bartlett Learning

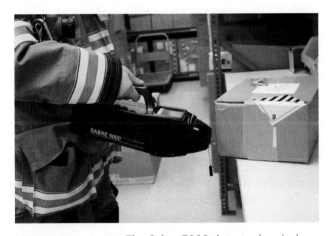

FIGURE 11-16 The Sabre 5000 detects chemical warfare agents, toxic industrial chemicals, and explosives.
Courtesy of Smiths Detection and Chris Hawley.

materials that would be used (**FIGURE 11-16**). The SABRE 5000 also covers materials that are likely to be involved in ordinary criminal activities. The device can also detect common explosive materials and common street drugs. It can detect vapors of some materials, while others, such as explosives and drugs, require a wipe test, which is placed into the device. For some reason, the first responder community has not latched on to the IMS technology or this device. It by far is the most versatile of the WMD detection devices and has the widest coverage. The false-positive issue remains even with this much-improved detector, as it is the nature of the technology. The software's algorithms are fairly tight,

and only chemicals that closely resemble the desired gases cause an alarm. For some devices, the problem is that the algorithms for the warfare agents are looking for indications of phosphorus and sulfur molecules in the chemical makeup of the sample gas. The detector alarms when near organophosphate pesticides, which is not bad, as the chemical structures are very similar to nerve agents. Other problematic chemicals are cleaning solutions, usually those for carpets and tile floors. They usually contain phosphorus or sulfur and have other components that mimic warfare agents. It is technically feasible to engineer out some of these false positives, but the tighter the algorithms become the more there is a chance that the unit could provide a false negative. A false negative means the detector did not alert to the presence of an agent when it should have. We can deal better with a false positive, and if we treat a false positive as a true event, there is less chance of harm than there is for a detector to say that the event is safe when it is not safe. There are too many variables in the types of warfare agent mixtures that a terrorist could come up with, and it is better to have some false positives than to risk missing an off-kilter nerve agent mixture that may not be picked up by a tight algorithm. The biggest problem with false positives mentioned previously comes with the detection of VX. Because VX has a very low vapor pressure, it is a challenge to get VX into the device, and it may have a wide algorithm. Some devices will have the ability to detect VX separately, which would be recommended as it limits the number of false positives.

Flame Photometry

This type of detection is comparable to the flame ionization discussed in Chapter 7. Like the flame ionization detector (FID), this type of detection relies on the use of a hydrogen flame to perform the analysis. The FID is used as a qualitative device and is quantitative using correction factors, and it provides an indication of the presence of a material. Flame spectroscopy, on the other hand, is qualitative and quantitative. It can determine the identity and amount of a contaminant. Its biggest disadvantage is the requirement to have a hydrogen gas cylinder, often available only from the manufacturer.

Flame spectroscopy is based on the emission of light spectra, commonly referred to as color spectrum. Each type of material, when ionized (excited), can emit photons. These photons (light) emit a specific wavelength when excited. As each element within the molecule provides a differing type of photon and therefore a specific color, the detection device can detect the type of compound. Flame photometry is an emission technique. When the element within the molecule is excited, it emits energy at specific emission lines. The manufacturer chooses which emission line to measure, generally chosen to minimize overlap/interference of emissions from other elements. What makes the detection device specific is the ability to identify the particular type of spectrum that the materials emit such as sulfur, nitrogen, phosphorous, and arsenic. This is done using filters, which "stop" some parts of the light spectrum, while allowing others to pass. The **AP2C** and the AP4C, manufactured by Proengin Incorporated, are designed to detect nerve, blister agents, blood agents, and a variety of toxic industrial chemicals (**FIGURE 11-17**). They use two optical filters that allow the light spectra for phosphorus and sulfur to proceed to a sensor. The intensity of the light helps to determine the concentration. Although the AP2C and the AP4C

are designed to detect chemical warfare agents, such as nerve and blister agents, they can detect many other compounds, based on the spectra developed. Proengin has other detection devices that use flame spectroscopy to detect other industrial chemicals, and their filtering systems vary depending on the compound. The AP4C can detect hundreds of common industrial materials.

Surface Acoustical Wave (SAW)

One technology that never caught on is **surface acoustic wave**, commonly referred to as SAW technology. It was one of the best technologies for the detection of chemical warfare agents, with the lowest number of false positives as compared to IMS and flame photometry. Its disadvantage is its relatively high detection threshold level; it cannot detect as low a level of chemical agent as IMS. The basic principle of a SAW sensor is that acoustic waves are generated and read by the sensor. The waves vary depending on the surface type and the material being either adsorbed or absorbed by the surface. The coating can be specific to a chemical, to a group of chemicals, or to any material that has density. A piezoelectric crystal is coated for chemical warfare agents. This crystal detects when a chemical is absorbed into the sensor coating. The electrically charged crystal generates an acoustic wave, and when a chemical warfare agent comes into the sensor housing, the wave changes. The sensor reading the wave detects the change and notes the wave signature. This signature is much like a fingerprint, which is compared to other fingerprints in the software library.

The inherent advantage is that SAW units check multiple sensors to confirm the presence of a suspect material. The SAW technology uses two to five sensors to detect chemical agents, and this use of redundant multiple sensors makes this an accurate detection device. It does return false positives but not nearly as many as the other technologies. The use of SAW sensors will become more widespread as the coatings determine the sensors' selectivity. As the array of sensors increases, the technology will be able to detect a wider variety of substances accurately. The sensors in current units are small, but those undergoing development are extremely small. Many sensors can be placed into a detection device, increasing accuracy.

Combination Technologies

So far, there have been some limitations in all chemical agent detection technologies. Until two or more of the existing technologies are combined into one device,

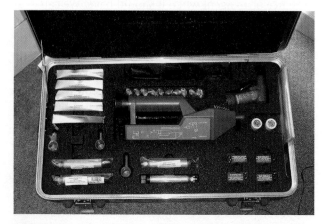

FIGURE 11-17 The AP2C, which is a French chemical agent detection device using flame spectrophotometry, can detect chemical warfare agents.
Courtesy of Christopher Hawley.

responders are best served by employing at least two if not all three of the current technologies. Given the choice of IMS, flame spectroscopy, or SAW technology, choose two of the three, and you have a high probability of accurate results.

Biological Testing Devices

In this category, several devices can detect the presence of biological threat agents. Some general broad screening tests and some sophisticated laboratory devices can be used in the street. This is the one area of WMD response that has created some controversy, much of it from a lack of knowledge and experience. With chemical agents, responders are more comfortable, and the signs and symptoms for the most part are immediate. With biological materials, there are still many unknowns, and the signs and symptoms can be delayed for a few days to a few weeks. In some cases, unseen biological materials can cause dramatic effects in humans. One of the major concerns with biological testing is false negatives, which can occur with many testing devices. The test results may be negative, but there may be fatal quantities of material present, which can be determined only through laboratory testing. The list of common biological threat agents referred to as biological selected agents or toxins (BAST) or abbreviated as the select agent list is contained in **TABLE 11-5**.

TABLE 11-5 Select Agents List*	
Category A Priority Pathogens—Greatest Threat	
Disease/Toxin	**Causative Agent/Source**
Anthrax	*Bacillus anthracis*
Botulism	*Clostridium botulinum* toxin
Plague	*Yersinia pestis*
Smallpox	Variola (or Orthopoxvirus)
Tularemia	*Francisella tularensis*
Viral hemorrhagic fevers	Arenaviruses (e.g., Lassa and Machupo), Filoviruses (e.g., Ebola and Marburg)
Category B Priority Pathogens	
Disease/Toxin	**Causative Agent/Source**
Brucellosis	*Brucella* species
Epsilon toxin	*Clostridium perfringens*
Food safety threats	(e.g., *Salmonella* species, *Escherichia coli* (*E. coli*) O157:H7, *Shigella*)
Glanders	*Burkholderia mallei*
Melioidosis	*Burkholderia pseudomallei*
Psittacosis	*Chlamydia psittaci*
Q fever	*Coxiella burnetii*
Ricin toxin	*Ricinus communis*

(continued)

TABLE 11-5 Select Agents List (*continued*)	
Category B Priority Pathogens	
Disease/Toxin	**Causative Agent/Source**
Staphylococcal enterotoxin B (SEB)	*Staphylococcus aureus*
Typhus fever	*Rickettsia prowazekii*
Viral encephalitis	Alphaviruses (e.g., Venezuelan equine encephalitis, eastern equine encephalitis, and western equine encephalitis)
Water safety threats	(e.g., *Vibrio cholera* and *Cryptosporidium parvum*)

*Adapted from Biodetection Technologies for First Responders: 2015 Edition.

© Jones & Bartlett Learning

Protein Screens

There are two types of **protein screens**, one broad based and the other specific. A broad-based system known as **BioCheck™**, manufactured by 2020 Gene-Systems, indicates the presence of proteins. The presence of a protein turns the protein solution purple. Protein is found in living materials, including pathogenic bacteria and toxins, such as anthrax and ricin. Any protein, including harmless ones, causes a positive test; this is why it is called a broad-spectrum test. To run the test, a specially treated swab is used to pick up the suspect material and to place it into a test tube, which holds protein test solution. A second vial holds a pH check solution (biological threat agents should be fairly neutral). An important note is that the test is valid only when testing visible materials. Responders cannot expect accurate results by performing a swipe test on a surface without visible material. The manufacturer provides that the detection limit for anthrax is 100,000 spores, which is a small but visible amount. In the event of a negative test, a control sample is supplied to ensure that the test was not compromised by interferents. A negative test result with this test does not mean that a dangerous biological material is not present. There may be enough material present to cause harm but not enough to be tested.

A more specific protein test, produced by Gen-Prime, is called **Prime Alert™**. With this system there are two tests: One for bacteria and another for toxins. The tests are specific for at least 13 biological threat agents, which differentiates Prime Alert from the broad-spectrum screens (**FIGURE 11-18**). The technology is known as novel fluorescent-based technology, and results are read by a fluorometer. To run the test, a swab is used to collect the sample material, and a nonfluorescent dye

FIGURE 11-18 The Bio-Check system is a broad-spectrum protein test for potential biological threat agents.
Courtesy of Christopher Hawley.

is added to the sample. The dye binds only to certain cellular components found in biological threat agents. When the dye attaches itself to these components, they change shape and become fluorescent. This fluorescence is visible only in the provided reader. The Prime Alert system uses a separate test for toxins, specifically ricin and botulinum toxins. The sample is mixed with a provided solution and placed into the receptacle of a handheld lateral flow assay, which is much like the handheld immunochromatographic assays described in the next section. A negative test can mean that there is no biological threat agent present or that it is present below the detection threshold. These tests are different from the broad-screen protein tests because they react only to specific biological threat agents, as opposed to just any protein. They are also broad enough that they test for a wide variety of biological threat agents, as opposed to just one specific agent.

Handheld Immunochromatographic Assay (HHA)

At the center of debate on biological detection are **handheld immunochromatographic assays (HHA)**, sometimes referred to as lateral flow assays. These assays are used to detect the presence of biological threat agents. Most responders drop the immunochromatographic portion of the name and call them **handheld assays**. These devices have been in use the longest by first responders, with varied success. The original HHAs were of low quality and not very reliable, compared to today's version. There are several on the market, and responders have a choice of brands. When purchasing this detection device, research is key. The tests for common biological threat agents look and operate similarly to a home pregnancy test, although they are higher quality.

The basis for these tests is simple. A sample is obtained using a swab (**FIGURE 11-19**). When you are done sampling, the swab is placed into a small container of buffer solution. The amount of material collected and placed into the buffer solution is key to a successful test. Some manufacturers' tests are very susceptible to being overwhelmed by the amount of sample material, and only small amounts should be picked up. The "more is better" assumption is false and usually leads to rejected test results. Some devices require the test strip be read by a reader while other tests, such as the Biological Agent Detection Device (BADD), do not require a reader (**FIGURE 11-20**). Readers, such as the one provided by **RAMP™**, let users know that they have overwhelmed the test and that the sample should be diluted (**FIGURE 11-21**). Other manufacturers may report only that the test was negative, which could be a false negative.

FIGURE 11-20 The Badd test, which is a handheld assay test for biological threat agents.
© Jones & Bartlett Learning. Photographed by Glenn E. Ellman.

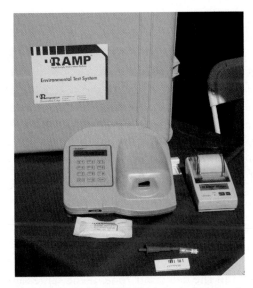

FIGURE 11-21 The RAMP biological test system, which has the ability to test for several biological threat agents, using handheld assays and a reader.
Courtesy of Christopher Hawley.

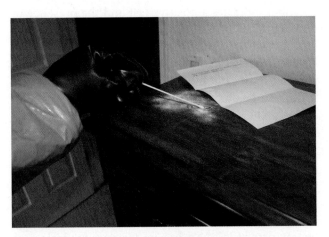

FIGURE 11-19 To collect biological material a small swab is used to collect a very little sample.
Courtesy of Rick Emery.

Once the swab has been placed into the buffer solution, a small amount of the buffer solution is then put into a test cartridge, which is specific to the threat agent. If you want to test for anthrax and the plague, you must run two tests, using the cartridge for each threat agent. Once the buffer solution is placed into the test cartridge, the liquid is drawn to the other end of it by following a membrane in the nitrocellulose carrier material. In the buffer solution are fluorescent-dyed latex particles that are coated with antigen-specific antibodies. If the sample has antigens, the antibodies bind, carrying the fluorescent particles. The sample is then carried to the detection zone inside the strip (**FIGURE 11-22**). The detection zone is coated with a second antibody, which is specific to the threat agent being detected. If the sample is made up of the antigens,

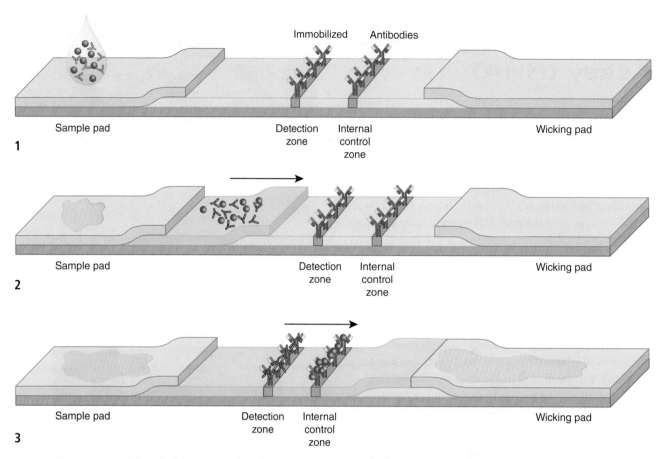

FIGURE 11-22 A handheld assay produced by Response Biomedical Corp. The RAMP system uses an internal control zone to ensure a good test. The sample is pipetted onto the sample pad as shown in (1). The sample material moves to the detection zone, as shown in (2). If the target agent is present, it attaches to the fluorescent particles, which are carried to the detection zone in (3). The reader detects these fluorescent particles and reports a positive hit. (Source: Response Biomedical.)

© Jones & Bartlett Learning

the report is positive, and the fluorescent particles are detained in the detection zone. If the target agent is not present, the fluorescent particles move beyond the detection zone. The reader is determining the presence of the fluorescent particles found in the buffer solution. Most tests run a confidence, or control, test to ensure that the device is functioning as it should. The RAMP system is easy to use and is representative of how the other manufacturers' tests are run. The internal portions of the RAMP system are unique, making it one of the more accurate tests. The test strip in most cases is placed into a reading device, which determines whether it is a positive or negative test. Always choose a system with a reader because without it the reliability of the testing drops dramatically. The Tetracore BTA Reader CX uses a reader that connects to your cell phone. The BTA Reader TX uses a tablet with a reader interface.

Another test uses an immunomagnetic sandwich assay that adds a few more steps but performs the testing in a shorter period. The sandwich assay has comparable fluorescent materials and a magnetic material. The threat agent (anthrax) is attracted to both these materials. If the target agent is anthrax, then it

binds with the fluorescent material and the magnetic material. Once they are combined, making a sandwich, a magnetic field is applied, which attracts the sandwich containing the anthrax and the fluorescent material. A wash step is added to remove any other contaminants and any excess fluorescent materials that did not bind. The presence of fluorescent material in the sensing chamber indicates a positive test and the presence of anthrax. The device has been tested by three independent laboratories and has been found to work as stated. The device has a stated detection threshold of <10,000 anthrax spores but in testing had the ability to detect lower levels and showed good consistency at 5000 spores and lower. As with many devices, the lower the threshold of detection, the greater the chance of false positives. The higher the threshold of detection, the chance for false positives is much less. The device is comparable to operate to its lateral flow assay cousins and has similar steps in sample prep.

Although ease of use is a good criterion for any detection device, it is of prime importance for biological testing devices. After the anthrax attacks in 2001–2002, most of the problems with street detection methods revolved around sampling methods and the failure to follow instructions. Comparing apples to apples is also good advice; make sure you try out each of the devices before deciding which one to purchase. Most of the biological threat agent technologies have been tested by one agency or another. Ask the manufacturers for the data from these test results, and read through the full length of the study. Check the manufacturers' claims against the test data, and see how accurate their claims are.

Polymerase Chain Reaction

Other than taking a sample and **culturing** a potential biological threat agent, a **polymerase chain reaction (PCR)** device is one of the best for identifying a material. The use of PCR devices is common in the laboratory and one of the first technologies used in the **Laboratory Response Network (LRN)** system (described later in this chapter). For street testing, there were only a few companies selling PCR units for first responder use. Now there is even a smaller number. The one that has been available for a long period of time and has a good reputation is the RAZOR EX, which is made by Bio-Fire (a bioMerieux company) (**FIGURE 11-23**). To understand how PCR units work, we must first explain the makeup of biological matter. All living matter contains **deoxyribonucleic acid (DNA)**, a nucleic acid from which genes are made. The structure of DNA is that of a double helix, and each strand of the double helix is a strand of DNA (**FIGURE 11-24**). These strands are made up of deoxyribose, a phosphate group (phosphoric acid), and a nitrogen-containing base. These bases are

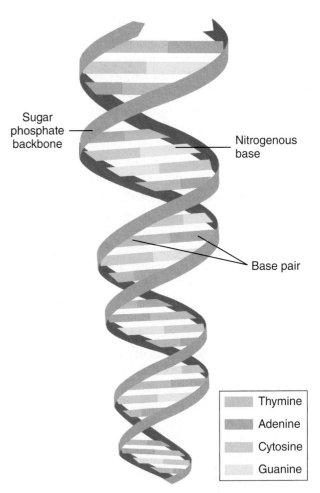

FIGURE 11-24 The structure of DNA and its basic components. DNA is made up of a sugar phosphate backbone and is held together by base and nitrogenous pairs (A, T, G, and C).
© Jones & Bartlett Learning

comprised of carbon, hydrogen, oxygen, and nitrogen. The bases are named adenine (A), thymine (T), cytosine (C), and guanine (G). The bases A, T, C, and G make up the rungs of the double helix ladder. The pairing of the bases is in a specific order—for example, A to T and C to G. The order of the bases creates the unique fingerprint and contains the genetic instructions for the organism. Once the sequence of the bases on one of the strands is identified, the sequence on the other strand is known. Within the gene there is a genetic code, which is made up of three nucleotides containing a combination of A, T, C, and G bases. The code is designated by the combination and order of the bases.

To determine whether the sample material is a threat agent, it must increase the amount of material to be read. To do that, there has to be an increase in the amount of DNA present. The PCR test artificially replicates (amplifies) the existing DNA so that it can be read. The PCR device unhooks the double helix and separates the strands. Free nucleotides can hook up to the exposed strands at these locations, which match the original order

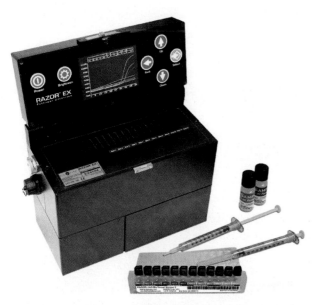

FIGURE 11-23 The Razor device is a PCR device to test for biological threat agents.
Provided by BioFire (a bioMerieux company).

of the nucleotides. If the wrong nucleotides hook up to the open strands, they are bumped off by the proper ones and cannot attach themselves. The PCR test uses a thermocycler to control the temperature, heating and cooling the sample as necessary. Added to the sample are specific primers and DNA polymerase, which are used to replicate the existing DNA. The primers allow only the desired section of DNA to be replicated and thereby amplified and subsequently measured. The heating and cooling cycles allow for the double helix to be separated, and the DNA polymerase fills in the newly created gaps. The polymerase copies the original DNA present in the sample, thus artificially increasing the amount of target DNA. The primer is made of fluorescent-labeled DNA and attaches itself to the sequence of the DNA matching the genetic fingerprint of the target. Thus, fluorescence increases as the amount of DNA that is being replicated increases. The PCR device looks for the increase in fluorescence, which happens only if a very specific genetic sequence or fingerprint of the target organism is present. For this reason, a PCR is a very specific testing process, especially in the laboratory. PCR can detect very small amounts of threat agents, although the lower the level of detection goes, the more cycles of heating and cooling are needed. PCR units can detect one anthrax spore, if you were willing to wait out the cycle time, but most responders balance time with spore count, with most units detecting less than 10 spores, well below the published fatal dose. The biggest challenge with PCR units is sample preparation, a concern with all biological threat agent testing. The easier the sample preparation is, the more likely the test will be successful. Manufacturers' devices differ in this area; some have easy sample preparation, with a minimum of steps, while others are complicated. A sample detection flow chart is provided in **FIGURE 11-25**. Compare the ease and number of steps, along with the cost of the disposable supplies, before you buy a PCR machine.

Explosive Detection Devices

There is some crossover with the chemical detection devices such as the SABRE 5000 and other IMS units. In addition, the FTIR and Raman units are beneficial in the field of explosives detection. In addition to these units, additional technologies have not been covered. FLIR has several models, Fido X2, X3, and X80. These devices use multichannel fluorescence technology (**FIGURE 11-26**). This technology is a series of polymer-coated strips, and if the vaporized sample causes fluorescence, then a hit is indicated. There are multiple sensing strips inside the device all designed for various explosives. The device can be used to swab people or other items to check for explosives (**FIGURE 11-27**).

The **E3500** device, manufactured by Autoclear (Scintrex Trace of Canada), uses chemical luminescence as its detection method. Nitrogen dioxide (NO_2) molecules emit infrared light when excited, which can be detected.

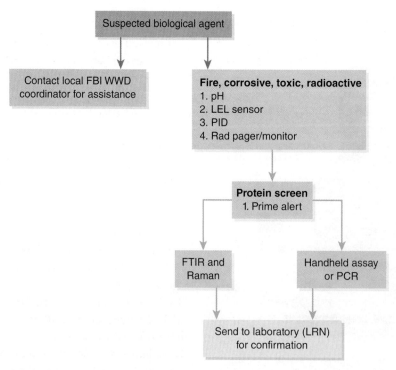

FIGURE 11-25 Sample biological detection flow chart.

FIGURE 11-26 The Fido X2 is an explosives detection device.
Provided by FLIR.

FIGURE 11-27 Test strip which can be used to wipe target areas.
Courtesy of Christopher Hawley.

In another method, this form of detection heats the sample material and looks for any nitrogen or nitrate groups. When explosive materials are heated, they produce nitrogen monoxide (NO), which mixes with ozone (O_3) in a chamber inside the detection device. With a red filter for the infrared light, a photomultiplier detects photons coming from the excited NO_2. A gas chromatograph (GC) can be used to determine the type of material present.

The E4500 uses gas chromatography-chemical luminescence (GC-C) technology. The E5000 device uses a combination of gas chromatography and IMS to identify explosive materials.

One of the most exciting upcoming technologies is **microelectromechanical sensor (MEMS)** technology. Developed by Oak Ridge National Laboratories (ORNL), and licensed to Sea Coast Science, this technology, also referred to as cantilever, is based on the use of very small **microcantilever sensors** placed on a circuit board. The sensors are coated with specific materials to attract or repel certain molecules. When the target molecules land on the sensor, the board is deflected, and it can detect the change in mass or stress on the sensor. By using an array of sensors specific to several different molecules, the detection device can be very specific. The Sea Coast unit combines this technology with mini gas chromatography. In addition to explosives detection, several companies are looking to use cantilever technology for chemical and biological detection.

The Electronic Sensor Technology Company has a unique nose-like device. The **EST-4200 Ultimate Vapor Tracer**, a combination SAW unit and gas chromatograph, provides a wide variety of detection possibilities. The device can detect drugs, chemical compounds, and explosives. It is marketed and designed to detect various odors and map them, using both sensing technologies. It requires helium as its carrier gas and is easy to operate, calibrated using a **BTEX** (benzene, toluene, ethyl benzene, and xylene) calibration standard. It can reach parts-per-billion levels in about 10 seconds, but its library is its limiting factor. Although many compounds are provided, it is not comparable to a stand-alone GC/MS device. The GC portion of the device separates the mixture into its individual chemical components; the SAW sensor determines the presence of each chemical as it exits the GC. The concentration can be determined by its interaction with the SAW sensors. An acoustic frequency is developed with the SAW sensor, which is one part of the fingerprint. The other part of the fingerprint is the travel time in the GC column. Combining the information regarding the interaction with the SAW sensor and the GC column develops a total fingerprint. Electronic Sensor calls the resulting SAW fingerprint VaporPrint™ Imaging and maps the image on a radial graph. The manufacturer states that the device mimics the human nose in the detection of various materials. The device can detect cork taint in wine, differences in soup, explosive materials, drugs, and hazardous materials. It has been tested against the usual explosives, drugs, and WMD materials.

The last device in this category, known as the **Sniffex** and manufactured by Sniffex Corporation, is included with some hesitation because it is very unusual and does not work. It was designed by a Yuri Markov, a Bulgarian engineer, and is in common use in Eastern Europe. The claims of its intended use and abilities come from the manufacturer, with a side note at the end. This pocket-sized handheld device is advertised as being able to detect not only the usual range of explosive materials but also explosive materials or a recently fired weapon at a distance of 328 feet (100 m), including those in buildings or vehicles. Detection is based on the presence of free radicals of nitrous oxide, along with the interaction of the Earth's crust, the device, and the user's body. The device must be held in a human hand and in a certain manner. The manufacturer says it can

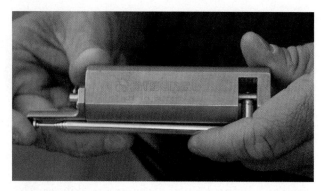

FIGURE 11-28 Sniffex is a handheld detection device that the manufacturer claims can detect explosives.
Courtesy of Christopher Hawley.

detect 1.7 ounces (50 g) of explosive material, behind a barrier, within 6.5–16.4 feet (2–5 m). It can also detect 35.2 ounces (1000 g) of explosive material behind a barrier at 32.8–65.6 feet (10–20 m).

Here is the interesting part: The device is essentially an explosives divining rod, similar to the sticks used to detect the presence of underground water sources (**FIGURE 11-28**). Testing has shown that this device is not reliable. The company was sued in 2008 by the Securities and Exchange Commission (SEC) for fraud. There are two other fraudulent devices like the Sniffex. They are manufactured by the U.K.-based Global Technical Ltd., which sold a device called a GT200, and U.K.-based ATSC, which sold the ADE 651, and are both the divining rod style. Another device called the Quadro Tracker from a South Carolina company was also selling these types of devices prior to the FBI arresting the owner.

Sampling Strategies for Terrorism Agents

Of all the tasks a hazardous materials team may get involved in, terrorism has great potential to be a literal minefield of concerns. There will be great pressure from several fronts to identify the potential agent and to assist in the mitigation of the event. There is potential for large numbers of patients, who may or may not be cooperative, and there may be considerable hysteria with the victims and responders. In the United States, there are about 7000 explosive recoveries and 700 actual bombings each year; the next greatest number of terrorism-related events are hoaxes. The emergency response to one of these hoaxes determines whether the terrorist is successful.

To be successful, the terrorist does not actually need to have sarin nerve agent but only needs to have you think he has it. How we respond to this potential

FBI AND CDC LABORATORY RESPONSE NETWORK

When discussing WMD detection, it is important to note the laboratory system that the FBI and Centers for Disease Control (CDC) have set up. The Laboratory Response Network, known as the LRN, handles the initial analysis of potential WMD materials. The LRN has been in existence since 1999 but really came to light in 2001–2002 after the anthrax attacks. The FBI partnered with the CDC to ensure that standardized protocols were implemented and that both public health and evidentiary concerns were addressed. The LRN system is still growing, and there are still limitations and/or problems with the type and amounts of testing that are done at the LRN level.

In each of the 56 FBI field offices, there is a WMD coordinator who is responsible for liaison with local responders. In the event of a potential WMD incident, the local responders should coordinate their efforts with law enforcement because potential evidence may need to be preserved. The WMD coordinator can assist in getting evidence or samples to an LRN facility.

The gold standard for analysis is laboratory testing, which is the desire of the FBI. Without laboratory testing, they cannot make their case, and the suspect may be allowed to go free. They do not place much emphasis on street detection methods and generally discourage messing with the potential evidence. Prior to shipment, they want the evidence screened for fire, corrosive, toxic, and radioactive hazards. The FBI has several hazardous materials response teams throughout the United States and maintains a highly technical response group known as the Technical Hazardous Response Unit (THRU), who can provide assistance when potential WMD materials are found or suspected.

There are more than 150 LRN labs set up for biological threat agents throughout the United States, and generally the local or state health departments run the labs. For chemical attacks, there are 62 laboratories that can conduct chemical threat agent analysis. LRN labs fall into three categories: National, reference, and sentinel. On the local level, hospitals serve as the sentinel labs, of which there are thousands. The reference labs, also known as confirmatory reference laboratories, handle the potential WMD materials. These are the labs that most responders send their samples to, through the FBI WMD coordinator. The national labs are used for highly infectious materials and have the highest testing capabilities. These laboratories follow a standard testing protocol and test only for a small number of threat agents. Unfortunately, this is a weakness in the system, and responders may need to obtain further analysis to determine the identity of a material.

threat directly determines who wins, and obviously we need to win this battle. When dealing with a potential terrorism event, we need to think as though we are at war, a unique and troubling concept.

Detection and protection is a concept that was co-developed with Frank Docimo for a National Fire Academy terrorism program. This concept is that detection is the first priority while wearing an appropriate level of protection. Responders need to wear appropriate levels of protection that take into account the chemical, the amount of chemical, the chemical risk category, heat stress, psychological stress, work task, and an overall risk/benefit analysis. In many cases, only the first item, the chemical, is considered, which endangers the life of the entry team by generally exceeding the actual type of PPE required to protect the responder. Safe but very rapid identification is necessary. The hazardous materials team must focus on identification, which can solve many of the issues. The hazardous materials team, as best it can, needs to ignore extraneous issues and focus solely on identification. Training of first responders is necessary to educate them in the fact that the hazardous materials team functions differently than they usually do, and first responders may need to pick up additional duties. The key to dealing with a hoax is to determine that the agent supposedly present is, in fact, not present. The first responders may be initiating mass decontamination procedures, and if there is no toxic or harmful material present, then decontamination is not necessary. If the stated agent is not present, or the unidentified material is not presenting any risk, the situation can be downgraded, and the community can return to somewhat normal conditions.

With the number of false positives that can be indicated, the detection of warfare agents is very difficult. A chemist who is familiar with hazardous materials emergency response is an important asset in determining the identity of the unknown material. For terrorism events, you can consult with the FBI's **Technical Hazards Response Unit** (THRU) through your local FBI WMD coordinator. The FBI THRU can link you up with its scientists and responders, who can provide recommendations for sampling strategies. Most FBI field offices have hazardous materials trained agents, and some of the larger FBI field offices have hazardous materials teams who have training and equipment for detecting warfare agents and other hazardous materials.

When doing any chemical sampling, it is important to follow the considerations given in this section. Each response has varied conditions, and each situation is unique. The sampling strategies are suggested starting points. A chemist or other technical specialist can assist in the sampling strategy, and the conditions present may dictate a change. Most important, when sampling for warfare agents, be careful of false positives, as they are common. Use a variety of tools in the identification of an agent. Do not rely on any one test to make an identification. Use biological indicators (the victims), test strips, and other detection devices. If they all indicate a nerve agent, then the material is most likely a nerve

agent. If the biological indicators are not suggesting any problems, and the M8 says it is nerve agent, then it is most likely not nerve agent. Even if more than one indicates nerve agent, it is most likely a false positive. If you use two differing technologies, such as a SABRE 5000 and an AP4C, and both indicate nerve agent, then it is most likely nerve agent, as they react to different false positives. Devices with the same technology will have some similarities with false positives, so that combination would not be accurate. The use of an FTIR or Raman is of great assistance in identifying a potential chemical warfare agent. The combination of these devices with an electronic vapor detection device, such as a SABRE 5000 or the ChemPro 100i, provides a sound confirmation about the presence of a chemical warfare agent. The basic rules that need to be followed while sampling any unidentified materials are the following:

- At a minimum, always wear respiratory protection.
- Avoid contact with the product.
- Sample for the risk category (fire, corrosive, toxic, and radiation), and always use a four-gas instrument, PID, and pH detection.
- A minimum of emergency decontamination must be available.
- Grab a sample, and do the testing away from the hazard area.
- ChemPro, FTIR, Raman, colorimetric tubes, M8, M9, and the M256A1 kit can be used to further characterize the unidentified material.
- Never guess as to the identity of the material until you have all the results.
- Back up street tests with lab tests, and follow chain of custody.
- Consult with a chemist.

In some cases, the detection devices for warfare agents have a minimum detection limit above the immediately dangerous to life or health (IDLH). By using several detection devices, you are more than adequately protected, and exposure values are conservative. The more accurate limits are provided in the lethal concentrations and incapacitating values. In all cases, standard and military detection devices can detect those substances at values well below those established for the warfare agents. **TABLE 11-6** provides a listing of the common agents and their various exposure levels, but the more realistic values are the LCt_{50} and the ICt_{50}. LCt_{50} is lethal concentration to time for 50 percent of the exposed population; in this case, the time usually means 3 minutes. ICt_{50} is incapacitating concentration to time for 50 percent of the exposed population. The time in this case is also unable to help themselves.

TABLE 11-6 Warfare Agent Exposure Values

Agent	Vapor Pressure (mmHg @ 25°C/77°F)	PEL (ppm)	IDLH (ppm)	LCt_{50} (ppm)	ICt_{50} (ppm)
Phosgene	1420	0.1	2	791	395
Chlorine	5830	1	30	6551	620
Sarin	2.86	0.000017	0.03	12	8
Tabun	0.07	0.000015	0.03	20–60	45
Soman	0.4	0.000004	0.008	9	4
VX	0.0007	0.0000009	0.0018	3	2
Mustard	0.11	0.0005	0.0005	231	30
Lewisite	0.58	0.00035	0.00035	141–177	<35
Hydrogen cyanide	742	10	45	3600	n/a
Cyanogen chloride	158	0.2	Unknown	4375	2784

© Jones & Bartlett Learning

Warfare Agents Sampling Strategy

The detection of warfare agents is based on military grade agent, something which is not likely found in the real world. With this in mind, a loose sampling strategy looking for various items should be adopted, such as those found in Chapter 12. In many cases, if the terrorists have developed the agent themselves, it will have impurities, which should increase the ability of some detection methods, such as colorimetric tubes. A homemade batch of warfare agent may not get picked up by other detection devices, though. The Aum Shinrikyo cult in Tokyo, Japan, used a 37 percent sarin mixture with acetonitrile in hopes of using the high vapor pressure of the acetonitrile to disperse the sarin, a flawed theory.

Some of the hydrolysis products for the nerve agents are common substances and can easily be detected using standard detection devices. The amount of hydrolysis products will vary with each agent but should be present in all but extremely pure substances. **VX** will always have hydrolysis products no matter how pure the substance is, which aids in its identification. **TABLE 11-7** provides a listing of agents and their hydrolysis products. **FIGURE 11-29** provides a sampling strategy when dealing with nerve agents. When dealing with VX, it may be best to pipette some of the liquid into a sample

TABLE 11-7 Hydrolysis Products of Warfare Agents

Agent	Hydrolysis Product
Sarin	Phosphonic acid and hydrofluoric acid. Sarin is also fluorinated, which also may be indicated.
Tabun	Phosphonic acid, hydrofluoric acid, and cyanides
Soman	Phosphonic acid and hydrofluoric acid
VX agent	Will always have thiol amine. Phosphonic acid may be present in early stages in a pure agent and will also be indicated by accompanied heat increase of the mixture.
Blister agents	Hydrochloric acid

© Jones & Bartlett Learning

jar and then heat the sample jar using a hot water bath to 150–160°F (66–71° C), then sample the vapor space with the detection devices. This method should be used with the other materials as well, especially if no readings were obtained during the first round of sampling.

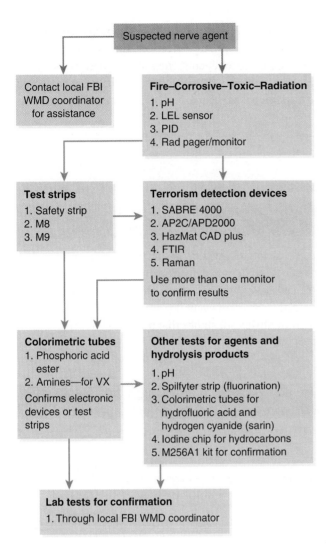

FIGURE 11-29 Sampling flow chart for nerve agents.
© Jones & Bartlett Learning

F paper or the Spilfyter strip is used predominantly to look for fluorine, but the petroleum hydrocarbons test may indicate other contaminants. The iodine chip test is a quick method of classifying unidentified hydrocarbons.

One of the immediate problems with the nerve agents is that their chemical structure is so close to organophosphate pesticides that an exact identification is nearly impossible in the field. The bottom line is, whether you have a military nerve agent or a commercial organophosphate pesticide, it does not matter. The treatment of victims, decontamination, mitigation, and PPE are all the same. The exact identification aids in the criminal aspect of the investigation, but that can wait for the lab results.

Blister Agent Detection

It is important for hazardous materials teams to check for blister agents quickly, as they have delayed effects, and victims may not be showing any signs or symptoms

after an exposure. Fortunately, with blister agents, there are not many common chemical equivalents, so when most of the detection devices indicate the presence of a blister agent, it is quite possible that the material is a blister agent. The one factor going against the detection of blister agents is the fact that their vapor pressures are extremely low. Much like VX, it may be best to heat the sample in a sample jar to see if that improves the off gassing of hydrolysis products. Some common chemicals will cause military detectors to indicate the presence of blister agents. The most common one is oil of wintergreen (methyl salicylate), which is used as a simulant for blister agent. Anything that has this flavoring in it will indicate on an electronic device. Mouthwash, breath mints, gum, and many candies have methyl salicylate as the wintergreen flavoring. In most cases, these items can be confirmed by other tests. If all street-level detection devices show blister agent, there is a pretty good chance the material is blister agent. The sampling strategy for blister agents is found in **FIGURE 11-30**.

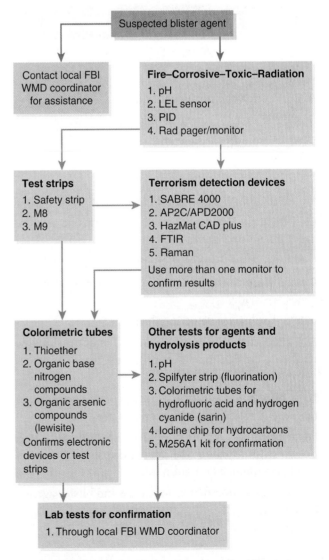

FIGURE 11-30 Sampling flow chart for blister agents.
© Jones & Bartlett Learning

One of the hydrolysis products for the blister agents is hydrochloric acid, for which there are colorimetric tubes. Some generic acid sampling tubes will also indicate by a distinctive color change for hydrochloric acid. The PID or FID must be close to the sample in order for it to read, and it can be anticipated that the readings will be low. Generally, if all military detection equipment indicates a positive for blister agents, then it is probably blister. An agent that provides false positives on all the devices would be difficult to manufacture, although certainly possible.

After-Action REVIEW

IN SUMMARY

The detection of terrorism agents is more difficult than a standard unidentified material. There are evidence issues, political issues, interagency issues, and potential for many victims counting on your characterization. Request help early, and request additional detection devices that can be of assistance. The key to survival in this or any other chemical incident is detection and adequate protection. You can save lives or reduce the hysteria caused by a potential terrorism event if you can quickly characterize an unidentified substance that may be a terrorism agent.

KEY TERMS

AP2C French chemical agent meter (CAM).

BioCheck Protein screen for biological materials.

BTEX Acronym for benzene, toluene, ethyl benzene, and xylene.

C2 Canadian agent detection kit.

ChemPro100i Warfare agent detection device.

Civil Defense Simultest (CDS) Colorimetric warfare agent detection system.

Culturing Process of growing biological material.

Deoxyribonucleic acid (DNA) Substance from which genes are made.

E3500 Explosives detection device.

EST-4200 Ultimate Vapor Tracer Detection device that can detect drugs, explosives, and chemical compounds.

False negative Detection device's failure to identify a target material when the material is present.

False positive Detection device's identification of a possible threat when one does not exist or incorrect identification of a contaminant.

GID-3 A detection device for nerve and blister agents.

Handheld assay See handheld immunochromatographic assay.

Handheld immunochromatographic assay (HHA) Detection device much like a home pregnancy test for biological materials.

Ion mobility See Ion mobility spectrometry.

Ion mobility spectrometry (IMS) A detection technology that measures the travel time of ionized gases down a specific travel path.

Laboratory Response Network (LRN) System of laboratories established by the CDC and the FBI for the identification of potential chemical or biological materials.

M256A1 kit A detection kit for nerve, blister, and choking agent vapors.

M8 paper A paper detection device like pH paper that detects liquid nerve, blister, and VX agent.

M9 paper A detection device used for liquid nerve and blister agents; comes in a tape form.

Microcantilever sensor Very small sensor with multiple coated levers that each react differently to groups of chemicals.

Microelectromechanical sensor (MEMS) Detection technology that uses cantilever sensors on a circuit board for the detection of a variety of materials.

Polymerase chain reaction (PCR) Biological detection device that examines the DNA of the suspect material.

Prime Alert Protein screen for biological materials.

Protein screen Test for material that contains protein, which could be a biological agent.

Raid-M-100 Warfare agent detection device.

RAMP Biological testing device.

SABRE 5000 Warfare agent detection device that detects explosives, drugs, chemical warfare agents, and toxic industrial materials.

Sniffex Explosives detection device.

Surface acoustic wave (SAW) A sensor technology used in detection devices, used in the SAW and hazardous materials CADS. After the sampled gas passes over the sensor, it outputs an algorithm that is checked for a possible match.

Technical Hazards Response Unit (THRU) Specialized response unit within the FBI for WMD and other high-hazard crime scenes.

VX A nerve agent, more toxic than sarin.

Review Questions

1. Which is worse—a false positive or a false negative?

2. What is the FBI's primary concern with a potential terrorism agent?

3. When collecting evidence, who is most likely to come in contact with a contaminated container?

4. M8 and M9 detect only which state of matter?

5. What is the biggest drawback of WMD colorimetric CDS tubes?

6. What is the most common electronic detection technology for WMD agents?

7. What provides the signature (fingerprint) of all living matter?

8. What is the gold standard in WMD evidence testing?

Tactical Use of Monitors

LEARNING OBJECTIVES

Upon completion of this chapter, you should be able to:

- Describe the use of a risk-based response
- Identify the methods that can be used to characterize an unidentified material
- Describe how detection devices respond to common materials
- Identify the role of detection devices in isolation and evacuation
- Identify sampling strategies for a variety of chemical hazards

Hazardous Materials Alarm

One evening, I was a couple of blocks from a fire station and noticed a medic unit headed toward a hospital. At the station, the emergency medical services (EMS) district supervisor told me I had missed a hazardous materials call. There had been a gas leak in the basement of the firehouse, and the hot water heater had exploded when the maintenance crew was relighting the pilot light. While watching the operation, he saw a big orange wall of flame pass by him and heard a "woof" sound. The only injuries were to the maintenance crew, and all were very lucky. The fire specialist on shift reported that, when told of the leak, the captain investigated it and stated, "It doesn't smell near the LEL [lower explosive limit]." This incident raises a couple of issues: First and foremost, the captain's nose missed the call because there was an explosion. Second, noses are not calibrated to anything, so it is impossible for a nose to quantify a little, a lot, or a boatload. Many chemicals, particularly sulfur compounds such as those made to odorize natural gas, tend to overwhelm the senses, and even though the gas is present, our noses shut off and no longer smell it. Responders should not use their noses to look for chemical odors, and they should be in full personal protective equipment (PPE), which includes self-contained breathing apparatus (SCBA). This situation could have had disastrous results, with a major explosion and a potential building collapse.

1. What risk category are you most likely to encounter in a chemical situation?

2. What level determines when there is extreme danger for the vapors to explode?

Introduction

The tactical use of monitors is by no means an easy skill, as there are many factors to consider, including the occupancy and location, the type of material and its state of matter, other identified chemical properties, spill location, weather, and finally the task to be completed (**FIGURES 12-1, 12-2,** and **12-3**). One of the challenges confronting response teams today is determining which suite of devices to start their monitoring strategy with. One of the best methods for learning how to effectively use monitors is to use them for every incident, even when the preponderance of information tells you that you really do not need all the instruments or that instruments may not be required at all. It is a good habit to carry monitors on calls that do not involve hazardous materials so you can learn what they do or do not react to. The monitors can be

FIGURE 12-1 When looking for hazardous gases and vapors responders should first look low as most gases/vapors are low to the ground.
© Jones & Bartlett Learning. Photographed by Glenn E. Ellman.

FIGURE 12-2 The next location to check is midway up the door.
© Jones & Bartlett Learning. Photographed by Glenn E. Ellman.

FIGURE 12-3 The last spot to check is up high, as only 11 gases go up.
© Jones & Bartlett Learning. Photographed by Glenn E. Ellman.

used when you do company tours or inspections, so you learn what you may expect to find in facilities under normal conditions. Knowing what kind of reactions and levels you can expect is invaluable when you arrive at a facility under emergency conditions. Some facilities have normal airborne concentrations of materials in the air. Suppose that the metal oxide detector in one chemical manufacturing facility reads 250 units. That number does not mean anything other than that it is 250 units out of 50,000 units, which could be considered a low reading. The problem is that an inexperienced responder on an emergency call may interpret 250 as part of the emergency condition and think that there is a problem when, in fact, there is no problem. A reading of higher than 250 for this plant would indicate a problem.

SAFETY TIP

This chapter on the tactical use of monitors provides some general rules and examples of tactical deployment of monitors. Each situation is different, and the responder should always wear a minimum level of protective clothing, and self-contained breathing apparatus should be the minimum. Successful monitoring is based on regular calibrations, bump tests, and training, but experience also is a factor. Science is not black and white; it has many shades of gray. It is the ability of monitors to detect these gray situations that will keep you safe.

SAFETY TIP

One of the best methods for learning how to effectively use monitors is to use them for every incident.

Risk-Based Response

The best use of monitors is in a **risk-based response (RBR)** profile, in which the monitors assist in the risk assessment of the situation. The monitors are your safety mechanism, much like a bulletproof vest that protects a law enforcement officer. The benefit of RBR is improved safety for responders and quick assessment, identification, and mitigation. By using improved monitoring skills, appropriate decisions can be made about PPE, isolation and evacuations, and the severity of the event. With RBR, these decisions can be made within minutes, usually less than 5 minutes for a well-trained crew. However, RBR requires educated and experienced personnel using monitors and interpreting the results. Monitors are dumb devices; they require human interpretation to make the right decision based on what the monitor reports. The first barrier that needs to be hurdled is that responders need to trust the instrument. This trust is based on the hope that the instrument is calibrated regularly, and the responder understands the true meaning of the interpreted results. A properly calibrated instrument does not lie; it is up to the responder to confirm the identity of the material and to interpret the numbers.

The basic premise behind RBR is the use of monitors to classify or characterize a chemical into one or more risk categories. The four risk categories are fire, corrosive, toxic, and radiation, and once a risk category is identified, crews can be properly protected. The RBR works well when dealing with an unidentified chemical or a potential release situation. Hazardous materials teams should avoid the use of the term "unknown," as it really does not exist. Hazardous materials teams do not respond to unknowns, as there is always something known that is available. People do not dial 911 just to get the hazardous materials team searching in some building looking for "something." There is a reason someone called 911, and that is known information.

LISTEN UP!

Four risk categories, fire, corrosive, toxic and radioactive, are the basic categories and encompass most risk materials. Explosives are considered fire risks. This list does not take into consideration other chemical and physical properties. We keep these basic categories to offer a quick response, as protective clothing and initial actions can be determined once we identify if a material is one or more of these categories. Even a group of chemicals that is mixed would fit into one or more of these categories. When quick action is required, then RBR applies. When time allows, further investigation can help determine a more finite risk.

Granted, the information may be limited, or the exact identification of the released substance may not be known, but every response has some "known" factors. Hazardous materials crews need to capitalize on those factors and use them for their benefit.

Using basic chemistry such as the state of matter assists us in the RBR process. A release of an unidentified solid material is not a major event, as a solid does not present an overt hazard unless you eat it, touch it, or apply energy. In the investigative phase, the level of PPE can be relatively low, as the potential for extensive contamination, then the level of PPE may be increased, but the tasks are different. No one chemical suit is best for hazardous materials situations. The risk category, chemical and physical properties, and the task determine the best type of PPE that should be worn. It is a misconception that a Level A ensemble offers the best or the highest amount of chemical protection. The hazards from the suit itself, such as heat stress, limited visibility, mobility, and communication, are all major life safety issues. Hazardous materials injuries are often from heat stress, not chemical exposures. There is great potential for fatalities if you believe that a Level A ensemble is the safest suit. Level A ensembles are appropriate for certain tasks, such as dealing with corrosive gases above the immediately dangerous to life or health (IDLH) level, or in toxic-by-skin-absorption IDLH atmospheres. A Level A ensemble is certainly appropriate when there is potential for whole body contact or when responders may become covered with a material that may be corrosive or toxic. No matter what the toxicity, a Level A ensemble is not appropriate for any situation in which there is a fire risk, and the flashover garments provide a false sense of protection. Flash garments are now designed to offer abrasion and some puncture resistance for chemical protective clothing, which increases their benefit. Remove the fire hazard, and then wear appropriate PPE. When dealing with a fire risk, wear appropriate fire protective clothing.

> ### LISTEN UP!
>
> The risk category, chemical and physical properties, and the task determine the best type of PPE that should be worn during an incident.

When using an RBR profile, the responder needs a minimum of five different detection technologies, which include pH paper, an LEL sensor, an oxygen sensor, a photoionization detector, and a radiation monitor (**FIGURE 12-4**). Using these five devices or sensors covers most fire, corrosive, toxic, or radiation risk categories. The LEL sensor (assumed to be part of a multiple-gas

FIGURE 12-4 Risk-based response requires pH paper, an LEL sensor, a photoionization detector, and a radiation monitor, and oxygen sensor a bucket makes it easy to carry all these devices.
© Jones & Bartlett Learning. Photographed by Glenn E. Ellman.

detection device) is typically housed with oxygen, carbon monoxide, and hydrogen sulfide sensors. The oxygen sensor is of prime importance because humans require oxygen, and the LEL sensor requires adequate oxygen to function accurately. When detecting the radiation risk, there are two options. Some prefer a radiation monitor, while others use a radiation pager. Either is acceptable, but see the discussion on radiation pagers in Chapter 9. The selection of protective clothing for hazardous materials can be controversial, and preferences depend on the identification of the risk category or categories of the situation. A potentially flammable environment probably requires firefighter protective clothing and SCBA to protect against fire hazards and inhaled toxic hazards. A Level B ensemble protects against the same inhalation hazards as firefighter protective clothing and against splashes from low-vapor corrosive materials. A Level A ensemble protects against inhaled toxic materials, high and low vapor pressure corrosives, and any other type of chemical exposure. Remember that Level A and B ensembles offer no protection against flame or heat. Potentially flammable environments should be avoided with any chemical protective clothing.

The selection of the appropriate protective clothing takes some thought, and no matter which type is chosen, response teams must train to quickly change the type of protective clothing when they encounter risks they are not protected against. RBR protocol, as part of a good response, includes anticipating the likely chemical hazards to be encountered so that appropriate PPE is selected from the start. Review your hazardous materials calls, and see which ones truly needed a Level A ensemble protective clothing. The HAZWOPER regulation (29 CFR 1910.120) mandates the use of a Level A ensemble in situations in which a material that is toxic by skin absorption is found above the IDLH level. The regulation also states that a Level B ensemble is acceptable for unidentified situations. The regulation does not mandate a level of protective clothing, except for the conditions provided for Level A ensemble. There are some conditions in which a Level A ensemble is recommended, which include confined spaces and highly toxic environments. A situation in which there are fatalities in a building is a probable candidate for a Level A ensemble, as well as situations in which extensive chemical contact is possible. Entering a building to investigate an unidentified powder is not a likely candidate for a Level A ensemble. We frequently encounter above IDLH conditions, including incidents with fatalities involving carbon monoxide, and we use firefighter protective clothing and SCBA, which is appropriate for the situation. Using a Level B ensemble for unidentified materials is recommended in HAZWOPER. One item to consider, though, is what is the risk of the commonly encountered chemical? Most hazardous materials teams respond to flammables and combustibles, usually in excess of 50 percent of the total number of responses. By choosing firefighter protective clothing and SCBA, one is protected against the chemical most likely to be found. In both the firefighter protective clothing and a Level B ensemble, SCBA protects against toxicity. The only true risk that a Level B ensemble does not protect against is high vapor pressure corrosive vapors, which are easily detected. Let the task dictate the level of protective clothing, and when coming in contact with a chemical, wear appropriate chemical protective clothing (CPC). When investigating a possible release, wear SCBA and firefighter protective clothing, which is more than adequate. There are very few chemicals that are toxic by skin absorption, and even those are usually found to be toxic at extremely high levels. One example is hydrogen cyanide, which is one of the most common toxic by skin absorption materials is not that toxic by skin absorption. Some studies have it being toxic at levels near 15,000 ppm and an exposure time of 20 minutes. The published IDLH for hydrogen cyanide is 50 ppm, which is considerably lower than the values that would create an issue through skin absorption. With that said, the OSHA regulations require the use of a Level A garment when the level exceeds 50 ppm.

Basic Characterization

Basic characterization of the event is done by the RBR system, and some tactical decisions can be made by the interpretation of these results. **TABLE 12-1** covers the top 10 most frequently released chemicals and what can be expected with the three detection devices. Understanding how these chemicals respond to the monitors puts you closer to a classification. The table is divided into several areas and provides some additional clues to identification. In many cases, there are no common materials that mimic the properties and readings of the chemicals listed.

Meter Response

This section provides some insight into how a variety of chemicals may affect the range of monitors. To classify an unidentified material, it must first be classified into one of the four risk categories. Once classified, if it meets only one of the four risk categories, responders can work to specifically identify the material through several methods. If it meets more than one risk category, then identification may be more difficult depending on the circumstances. Although it is not often mentioned in these sampling protocols, make sure to always determine the level of background radiation.

Corrosive Risk

Of all the risk categories, the corrosive risk is the easiest to determine. The detection device is easy to use and to interpret. The range of corrosives on the street is limited to a few common materials. If using multirange pH paper, a color change to red indicates an acid being present. If the pH paper changes above or away from the spill, then the acid is a high vapor pressure material. The most common materials that are acidic and have a high vapor pressure are hydrochloric acid, hydrofluoric (HF) acid, acetic acid, and oleum. HF acid has a more characteristic vapor cloud and is more aggressive than hydrochloric acid. F paper will indicate for HF but not HCl. Oleum is the most aggressive as a vapor cloud and reaction with other materials but does not indicate on the pH paper until you get fairly close to the spill. The higher the percentage of oleum, the farther away it will indicate. Many acids are low vapor pressure acids and only indicate when the pH paper is dipped into the liquid. Further tests are needed to narrow down the type of acid released. Common acids that are low in vapor pressure include sulfuric and phosphoric. They may be low in vapor pressure, but they are certainly a risk due

TABLE 12-1 Top Ten Chemicals Released and Meter Response

Meter Response	Ammonia	Sulfur Dioxide	Chlorine	Hydrochloric Acid	Propane
pH	>12, indicates in vapor; has a high VP	<2, indicates in vapor; has a high VP	<2, indicates in vapor; has a high VP	<1, indicates in vapor; has a high VP	No response
LEL	High levels in a building indicate a flammable atmosphere	No response	No response	No response	Yes
PID	Good response; a good detection device for ammonia	No response	No response w/10.6, but response with 11.7 eV lamp	No response	Only for 11.7 eV lamp
O₂	>5000 ppm will drop O₂ by 0.1% for each 5000 ppm present	>5000 ppm will drop O₂ by 0.1% for each 5000 ppm present	>5000 ppm will drop O₂ by 0.1% for each 5000 ppm present	>5000 ppm will drop O₂ by 0.1% for each 5000 ppm present	>5000 ppm will drop O₂ by 0.1% for each 5000 ppm present
CO	No response	No response	No response	Usually no response may indicate on some instruments; usually accompanied by a H₂S reading	No response
H₂S	No response	No response	No response	Usually no response may indicate on some instruments; usually accompanied by a CO reading	No response
Odor	Strong, irritating, will be noticeable	Strong, irritating, will be noticeable	Strong, irritating, will be noticeable	Strong, irritating, will be noticeable	Distinctive odor
Next step in detection	Colorimetric tube or FTIR to confirm	Colorimetric or FTIR	Spilfyter strip (liquids) or colorimetric tube to confirm	Colorimetric or FTIR	Colorimetric tubes or FTIR (gas), GC/MS

(continued)

TABLE 12-1 Top Ten Chemicals Released and Meter Response (continued)

Meter Response	Sodium Hydroxide	Sulfuric Acid	Gasoline	Flammable Liquids	Combustible Liquids
pH	>12, indicates only in the liquid	<1, only in the liquid	Neutral (check leading edge)	Usually neutral (check leading edge)	Usually neutral (check leading edge)
LEL	No response	No response	Yes	Yes	No response
PID	No response	No response	Yes	Usually yes, usually a quick response	Responds with low readings, must be close to the liquid
O_2	No response	>5000 ppm will drop O_2 by 0.1% for each 5000 ppm present	>5000 ppm will drop O_2 by 0.1% for each 5000 ppm present	>5000 ppm will drop O_2 0.1% for each 5000 ppm present	>5000 ppm will drop O_2 by 0.1% for each 5000 ppm present
CO	No response	No response	No response	No response	No response
H_2S	No response	No response	No response	No response	No response
Odor	No odor	No odor	Distinctive odor	Petroleum odor	Weak odor of petroleum
Next step in detection	Colorimetrics	Colorimetrics or FTIR (96%)	Colorimetric tubes, iodine chip, HazCat, FTIR, GC/MS	Colorimetric, iodine chip, HazCat, FTIR, Raman, GC/MS	Colorimetric, iodine chip, HazCat, FTIR, Raman, GC/MS

to their corrosiveness. Phosphoric is commonly found, as it is a component of soft drinks. The concentration of the material is another factor to consider when determining the risk a material presents.

Fire Risk

The catalytic bead LEL sensor only picks up flammable gases and those materials that are emitting flammable vapors. If a catalytic bead LEL sensor provides a reading of 1, then there is a flammable material released. If you cannot locate a liquid spill, then the reading is from a gas. If you find a liquid spill and the catalytic bead LEL sensor is reading, and as you get closer to the release the meter starts to climb, then the spilled material is flammable. The amount of the reading helps determine how the meter reacts to the material. The quicker and higher the meter goes, the more flammable the material is. A metal oxide sensor (MOS) detects both flammable and combustible materials. The MOS also picks up inorganic materials that a photoionization detector may not. Keep in mind that the MOS may also pick up some particulates in the air as well. The benefit of an MOS is that it detects tiny amounts of many things, long before they become harmful, and in some cases the other detection devices may not pick up the material. A material that causes an MOS to rise rapidly and go above 1000 is a flammable. The higher the number and the quicker the rise, the more flammable the material is.

Toxic Risk

The photoionization detector picks up organics and some inorganics. The most common material that a PID reacts very well to is solvents, most of which are flammable liquids. The quicker and the higher the rise, the more flammable the material is. The PID is used to determine the presence of toxic risks, but many toxic risks also present a fire risk as well. Once a PID goes above 500, the material is usually picked up by a catalytic bead LEL sensor. The MOS LEL tracks much like the PID and rises accordingly. However, the MOS cannot replace a PID, as the MOS technology is way too erratic to be depended on for definitive toxicity readings. It does fine indicating the presence of a material that should not be in the air. It cannot be related to parts per million and should not be used for this purpose. Once a PID picks up a material, you can use colorimetric tubes or the GasID for a possible identification. For liquids, you can use test strips, FTIR, Raman, or the HazCat kit. The FID could be used to determine whether the material is organic.

A common question is, what happens when a truckload of chemicals overturns, and they all mix together? The mixing of the materials is not a large concern unless they react with each other. Violent reactions usually occur prior to the arrival of responders, and any major event of this type is usually preceded by sufficient warning. When using RBR philosophy it really does not matter what the mixture is. When chemicals mix, they still may present fire, corrosive, toxic, or radiation risks. Using standard RBR, responders can be protected against all four hazards and can determine appropriate levels of protective clothing, isolation distances, and severity of the event. The exact nature of the mixture can be determined later and usually requires the use of an FTIR, Raman, or GC/MS device. Consulting with a chemist and a laboratory to assist would be helpful in identifying the exact nature of the mixture.

The Role of Monitors in Isolation and Evacuations

The use of detection devices enables the incident commander (IC) to make immediate decisions regarding isolation and evacuation strategies. Following the recommendations in some reference texts or plume dispersion models in most cases results in too large an area being evacuated unnecessarily. The distances used in references are very conservative and are set for a "normal" (worst-case) situation. By using the wide-ranging devices such as the LEL sensor and the PID, some initial isolation distances can be set. When readings begin to be encountered, this area becomes a potential isolation area (**FIGURE 12-5**). Isolation areas should be designated by hazard area, such as toxic, corrosive, flammable, explosive, collapse, and others. The designation should include the type of hazard that could be found in that area. If chemical warfare agents are probable or suspected, other devices such as the LCD or the ChemPro could be used for the same purpose (**FIGURE 12-6**). For normal chemical releases, the use of this tactical objective is easy, as the most common chemicals are easily detected. Using chlorine as the example, it would be easy to set up an isolation zone for a chlorine release. Some teams use the threshold limit value (TLV) or permissible exposure limit (PEL) or half that value to establish an isolation area. The IDLH would indicate the immediate hazard area, usually called the hot zone. All of the areas would be expanded to allow for wind changes and product movement, but real-time, on-site monitoring establishes the true conditions.

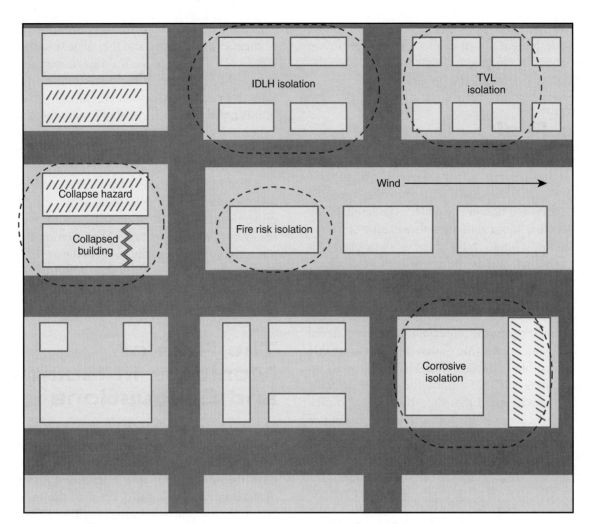

FIGURE 12-5 Isolation areas should be designated by hazard area, just not hot, warm, and cold.
© Jones & Bartlett Learning

FIGURE 12-6 If chemical warfare agents are suspected then a WMD detection device should be used.
© Jones & Bartlett Learning. Photographed by Glenn E. Ellman.

In some areas, especially major metropolitan districts, it is nearly impossible to affect an immediate evacuation. A high-rise building may have thousands of people working in it. A several-block area may have hundreds of thousands of people in it working and living there. To say that we will evacuate this area quickly and safely is living in a vacuum. It is better to determine the hazardous environment and isolate only those areas that are affected or could become affected due to a shift in weather conditions or other change.

Several systems network existing detection devices. The devices are usually an LEL sensor and a carbon monoxide, a hydrogen sulfide, and a photoionization detector. Some advanced chemical warfare agent detection devices can even be added to this network. These devices use either radio telemetry or wireless connections, while others are hardwired to a control point (**FIGURE 12-7**). The detection devices are monitored by a remote computer, and plumes can be measured and monitored. Most computer plume models that are incorporated into emergency response software, such as CAMEO, use projections based on inputted data. Using a variety of instruments downwind of a release and transferring this data to the plume projection software provide an actual plume model, one based on facts, not extrapolation.

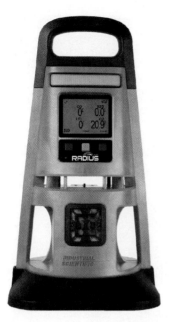

FIGURE 12-7 This detection device is part of a wireless network, and the detector readings can be tracked at a central location, such as the command post.
Courtesy of Industrial Scientific.

Detection of Unidentified Materials and Sampling Priorities

FIGURES 12-8 through **12-14** are flow charts to guide you through the process of characterizing an unidentified material. These figures are an attempt to provide a starting point, upon which you need to expand. Local conditions dictate which methods may work best, and the flow charts assume that all the devices mentioned therein are available to responders. Some initial cautions must be provided to this section, as a 100-percent-positive identification is nearly impossible in the field and requires lab analysis. However, you can characterize an unidentified material to a minimum of a risk category. Many of the systems in place today to characterize a substance rely on only one detection technology and do not consider other technology that may be available to a hazardous materials team.

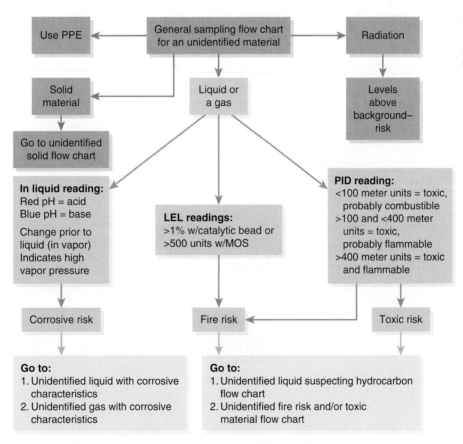

FIGURE 12-8 General sampling flow chart, to be used as the starting point.
© Jones & Bartlett Learning

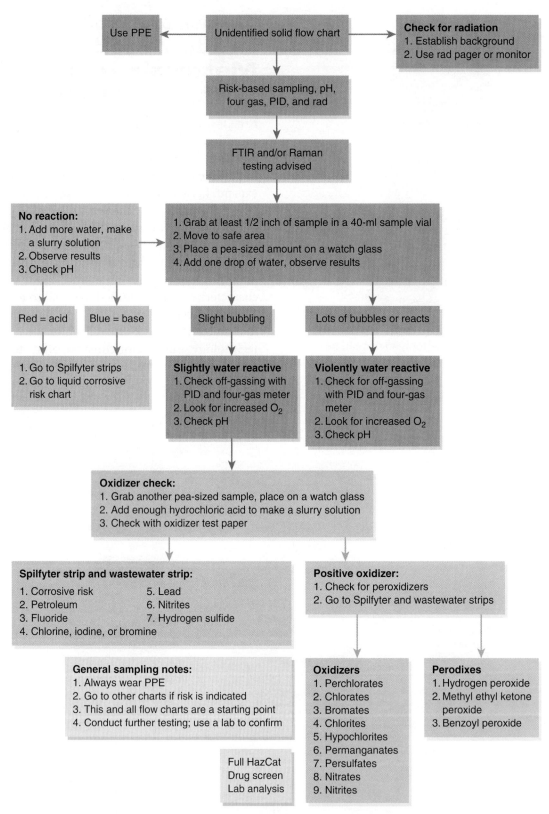

FIGURE 12-9 Sampling flow chart for unidentified solid materials.
© Jones & Bartlett Learning

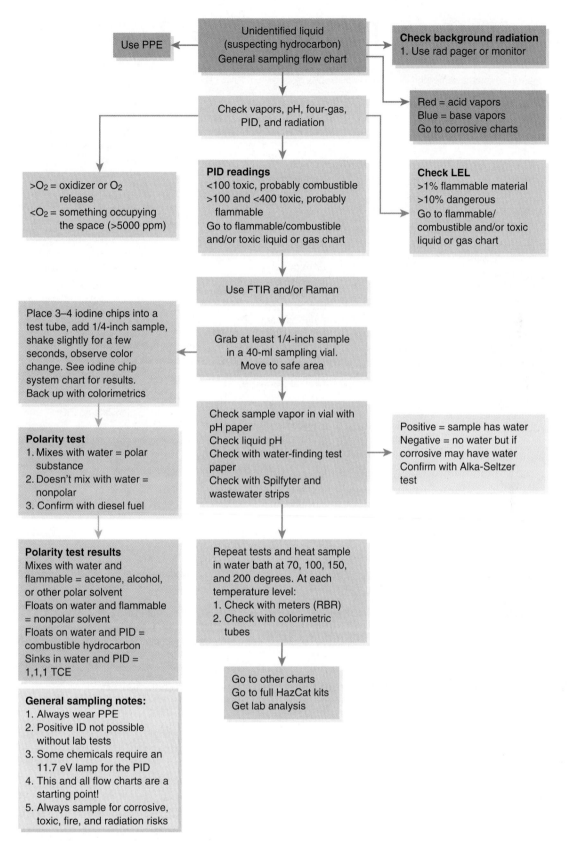

FIGURE 12-10 Sampling flow chart for unidentified liquid suspected of being a hydrocarbon.
© Jones & Bartlett Learning

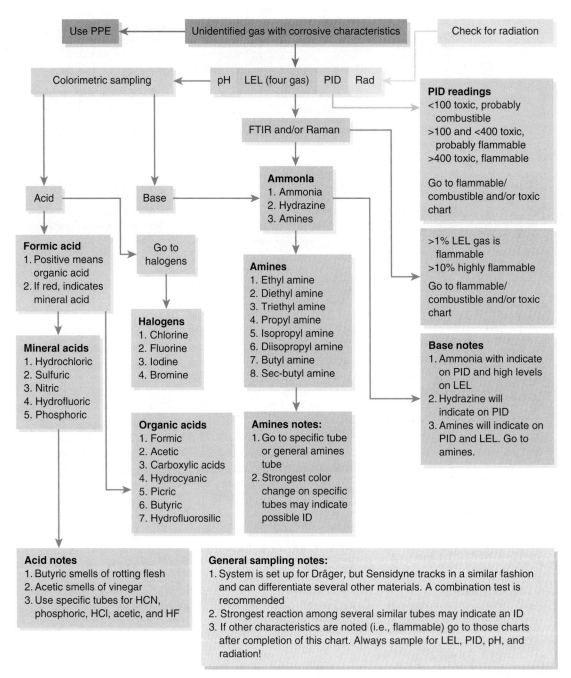

FIGURE 12-11 Sampling flow chart for an unidentified liquid with corrosive characteristics.

© Jones & Bartlett Learning

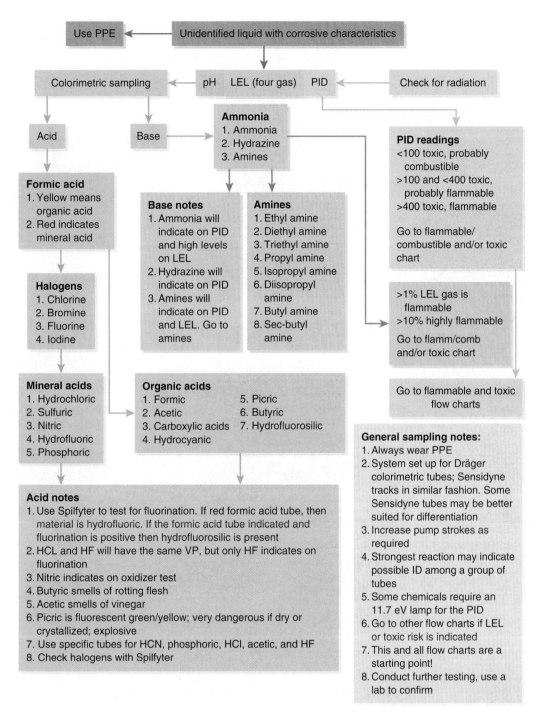

FIGURE 12-12 Sampling flow chart for an unidentified gas with corrosive characteristics.
© Jones & Bartlett Learning

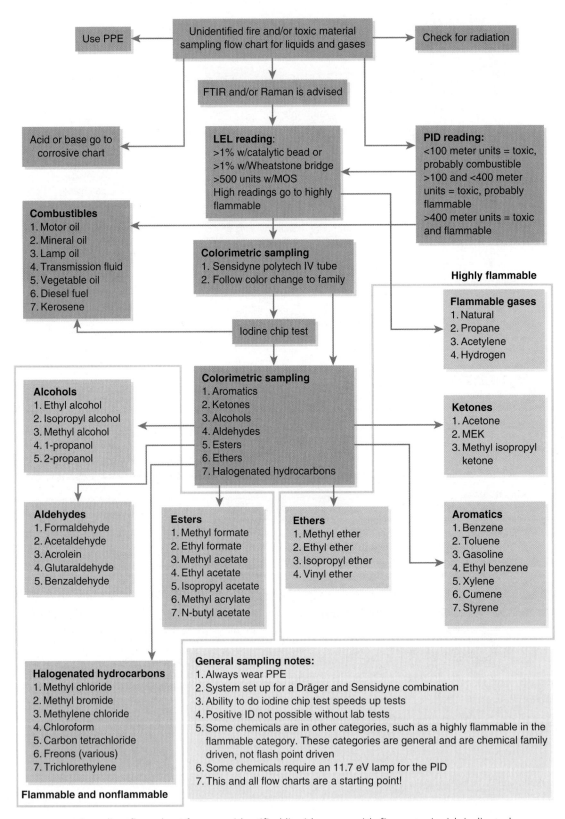

FIGURE 12-13 Sampling flow chart for an unidentified liquid or gas with fire or toxic risk indicated.

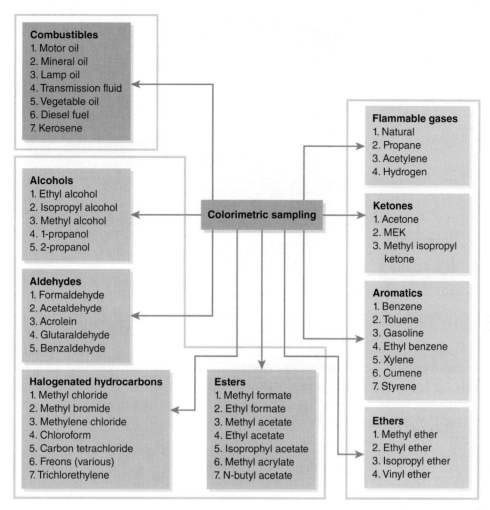

FIGURE 12-13 *Continued*
© Jones & Bartlett Learning

As with many things in the hazardous materials world, this information is what is normally found, and local conditions may change the way a chemical reacts, with weather being the predominant factor. One would not expect a chemical to act the same in Chicago in January as it would in Miami. The standard is 68°F (20°C), and you will need to calculate to any temperature other than that. Once you gain some experience you may be able to skip some steps in these flow charts, as street experience plays a role in characterization.

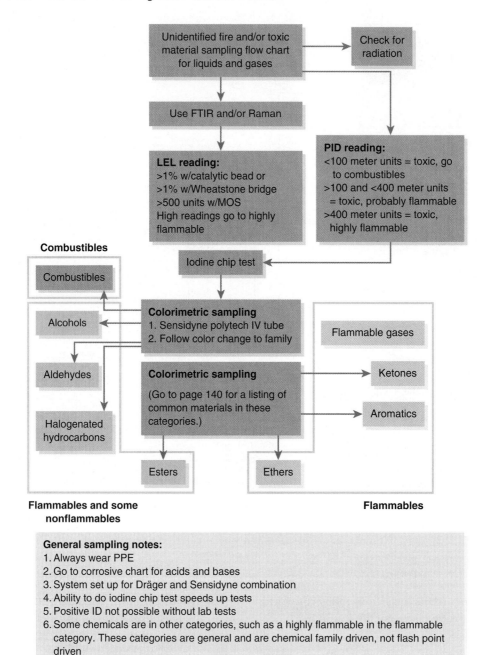

FIGURE 12-14 Flow chart for unidentified liquids and gases that may present fire and/or toxic risks.
© Jones & Bartlett Learning

After-Action REVIEW

IN SUMMARY

The idea of risk-based response is one that benefits the responders, the community, and anyone else who may be affected by a chemical release. Based on chemical characterization, responders can gain valuable response information safely and quickly. The whole concept of identifying risk, evaluating the meter's response, and determining isolation and evacuation areas are important skills for the responder to learn. As that could best be accomplished in written form, there are guides provided to the reader to help with the evaluation process. Keep in mind these are guides, and your own knowledge and experiences play a role in identifying an unknown material.

KEY TERM

Risk-based response A response methodology that characterizes risk.

Review Questions

1. What are the five detectors that should be taken into every potentially hazardous situation?

2. What risk category does most response teams respond to at least 50 percent of the time?

3. Will chemical protective clothing protect you against flame impingement?

4. What is the only way to identify a chemical with 100 percent accuracy?

5. When pH paper turns red this indicates what type of material?

6. An LEL sensor indicates with a reading of 1 percent; what type of material are you dealing with?

7. What is always the minimum level of protection that is required?

8. When is a Level A ensemble a good choice for protective clothing?

CHAPTER 13

Detection of Toxic Gases in Fire Smoke

OUTLINE

LEARNING OBJECTIVES

Upon completion of this chapter, you should be able to:

- Identify the toxic gases that are present in fires
- Identify the hazards of the toxic twins
- Describe the application of exposure levels in fire situations
- Describe the methods to identify toxic fire gases

Note: Much of this chapter is an excerpt used with permission from an article *Detection of Toxic Gases in Fire Smoke,* written by this author for the Fire Smoke Coalition. For more information on this and other fire smoke–related topics, please refer to the Fire Smoke Coalition, which is a great resource.

Hazardous Materials Alarm

Fire fighters have long understood the dangers of fire smoke as it is taught in firefighting classes, but even with this knowledge, there is a disconnect. Even fire fighters assigned to hazardous materials response teams have had a hard time making the jump to understanding or realizing the dangers. One of the biggest misconceptions is that only when you have visible smoke is when you are in danger. An engine company responded to a kitchen fire, in which a Teflon pan overheated causing a smoke condition. Fire fighters started developing headaches, and the hazardous materials unit was requested, where they found high levels of CO and HCN. Because the fire was out, and it was only light smoke, no self-contained breathing apparatus (SCBA) was used. In another situation, an engine company responded to a dumpster fire. The officer wanted to try out the new hydrogen cyanide (HCN) detector. He was upwind by about 30 feet (10 meters) and found toxic levels of HCN.

1. What color are most toxic gases?

2. What is the most common route of toxic gases?

3. What are the most common gases found in fires?

Introduction

It does not take long to read through much critical examination of the scientific data to determine **fire smoke** is a toxic soup of dangerous gases (**FIGURE 13-1**). Within all the significant dangers of fire smoke,

FIGURE 13-1 Fire smoke is a toxic soup of hazardous materials.
Courtesy of Rob Schnepp.

common dangers are the **toxic twins:** Hydrogen cyanide (HCN) and carbon monoxide (CO). What is still confusing for responders is how these gases are treated depending on the various situations in which they may be encountered. The dangers of fire smoke can often be under appreciated in today's response community (**FIGURE 13-2**). Think about it this way: If any *one* of these gases were leaking from a tank truck or cylinder on a highway, the first responders (fire fighters) would establish a perimeter, and only personnel in hazardous materials suits would be allowed in the hazard area (**FIGURE 13-3**). Now combine all these dangerous gases in a high-temperature environment and add them to a house fire where fire fighters will be running towards the situation—sometimes with and sometimes without proper protective clothing. Structural fires, vehicle fires, and, frankly, any type of fire will produce toxic gases—it is the toxicity and the quantity that varies from fire to fire. Something as simple as heating certain items, such as plastics, for example, can produce toxic gases. If you are a fire fighter who has a few years of experience, I am sure that at some point in your service you suffered a headache following a fire, which is an indication of an exposure to one or more of the toxic gases. Contents do not need to be on fire but instead merely exposed to the radiant heat to begin off gassing. These exposures can add up over a fire fighter's service and create long-term health issues. It is critical to the safety of our emergency response

FIGURE 13-2 Fire fighters in smoke should always wear SCBA. When fire fighters are in visible smoke particles they recognize the hazards. It is when they can't see the smoke that they forget that the toxic gases are colorless and sometimes odorless.

Courtesy of Rob Schnepp.

FIGURE 13-3 Responders would treat an incident with this truck very carefully and make sure they would wear appropriate protective clothing. We could easily put these same warning placards on a house fire.

Courtesy of Rob Schnepp.

community to determine what dangers are present and at what level they exist at the incident scene. This can present a significant challenge for the emergency response community, primarily considering that current technologies are not able to accurately distinguish between the components of the toxic soup known as fire gases.

Exposure Levels

The toxic mixture contained in fire smoke is comprised of thousands of materials but primarily consists of carbon monoxide, carbon dioxide, hydrogen cyanide, ammonia, hydrogen chloride, sulfur dioxide, hydrogen sulfide, and oxides of nitrogen. The National Institute for Occupational Safety and Health (NIOSH) establishes safe levels for chemical exposures in the workplace, which were discussed in Chapter 2. The values for the toxic gases commonly present in fire smoke are listed in **TABLE 13-1**.

There are several methods to detect toxic fire gases, but the most common methods involve the use of a single gas or multiple gas detection device. When monitoring for toxic fire gases, responders would use the levels listed in Table 13-1 to determine what is safe or not safe. The immediately dangerous to life or health (IDLH) for CO is 1200 ppm; therefore, if responders obtain meter readings that indicate levels at or above 1200 ppm, the environment or situation should be considered immediately dangerous to the life and health of those people in the environment or situation.

The use of the term "immediately dangerous" in IDLH should impart a sense of risk, but what gets a bit more difficult to navigate are when the situations where levels are below the IDLH. The recommended exposure limit (REL) is the most conservative of the remaining levels, and therefore safer, so we will use it for our examples. The REL is an average over a 10-hour period, so to get an REL on an emergency scene, it would involve calculations of the dose of the toxic gas over time. Considering that time tracking isn't typically available for emergency responders, and the dynamic nature of fire scenes, this isn't a realistic expectation. The simplest and safest method is to use the actual REL level itself as the action level. For example, the REL for CO is 35 ppm, so any level above 35 ppm would be considered dangerous and thus require responders to remain in PPE.

During hazardous materials responses, responders should not consider a building or area safe until monitored levels fall below the REL. It is recommended that at levels below the REL breathing apparatus masks may be removed, and at levels above the REL, breathing apparatus should continue to be utilized. Long-term exposure to toxic gases can be detrimental to your health, so the best practice is to wear your mask anytime toxic gases are present. At your first fire, you may not anticipate much exposure, but if you catch a second fire, you are adding to your exposure and increasing your risk and, therefore, adding to the risk at every subsequent fire during your service. The common-sense approach is simply this: Why inhale any toxic gases? It just is not worth the risk to you, your health, your livelihood, and your family.

TABLE 13-1 Exposure Values and Density				
Gas	REL	STEL	IDLH	Density
Ammonia (NH_3)	25 ppm	35 ppm	300 ppm	Lighter than air
Carbon dioxide (CO_2)	5000 ppm (0.5% vol.)	30,000 ppm (3% vol.)	40,000 ppm (4% vol.)	Heavier than air
Carbon monoxide (CO)	35 ppm—no higher than 200 ppm allowed (ceiling)	NR	1200 ppm	Lighter than air
Hydrogen chloride (HCl)	No higher than 5 ppm (ceiling)	NR	50 ppm	Heavier than air
Hydrogen cyanide (HCN)	4.7 ppm (15 minutes only)	4.7 ppm—1 time only	50 ppm	Lighter than air
Hydrogen sulfide (H_2S)	10 ppm (10 minutes only)	10 ppm—1 time for 10 minutes	100 ppm	Heavier than air
Oxides of nitrogen (NO_X)(NO_2) (NO)	NO_2—3 ppm NO—25 ppm	NO_2—5 ppm NO—NR	NO_2—20 ppm NO—100 ppm	Heavier than air
Sulfur dioxide (SO_2)	2 ppm	5 ppm	100 ppm	Heavier than air

© Jones & Bartlett Learning

Detecting Toxic Gases

Part of the challenge in detecting toxic fire gases relates directly to the volatility of a given situation and the limits of technology (**FIGURE 13-4**). The atmosphere is not consistent, and the type and amount of gases present will vary. In most cases on a fire scene, the toxic gases common in fire smoke will be present, but the levels may be below the threshold of detection. There are eight common toxic gases that are usually present, but many fire departments only try to detect CO and/or HCN. Responders that are only looking for CO could be missing at least seven other gases, and some are significantly *more* toxic than the CO! It is a certainty that there are more than the common eight gases present, depending on what is burning or smoldering. Using detection methods to search for CO and HCN may offer some a false sense of protection as there may be other silent killers present. While looking for CO and HCN does offer some level of protection and can be a good first step, truly effective detection and monitoring would mean you that responders would check for all eight gases that are most likely present.

The carbon monoxide sensor will react to quite a few other gases, some of which are listed in **TABLE 13-2**.

FIGURE 13-4 A single gas detection device looking for hydrogen cyanide in the smoke.
Courtesy of Christopher Hawley.

Typical electrochemical sensors used by fire fighters include oxygen, hydrogen cyanide, carbon monoxide, hydrogen sulfide, ammonia, chlorine, and ammonia. These sensors are commonly found in single-gas and

TABLE 13-2 Common Interfering Gases		
Common Interfering Gases to CO and H₂S Electrochemical Sensors		
Acetylene	Hydrogen	Methyl sulfide
Ammonia	Hydrogen cyanide	Nitric oxide
Carbon disulfide	Hydrogen sulfide	Nitrogen dioxide
Dimethyl sulfide	Isobutylene	Phosphine
Ethyl alcohol	Isopropyl alcohol	Propane
Ethyl sulfide	Mercaptans	Sulfur dioxide
Ethylene	Methyl alcohol	Turpentine

© Jones & Bartlett Learning

multiple-gas devices. There are a couple of issues that arise with electrochemical sensors, specifically those used to detect O_2, CO, H_2S, and HCN, which are detailed below:

- Toxic sensors react to other gases, and you cannot be sure you are reading a level of the intended gas or an interfering gas. For example, if you are reading levels of both CO and H_2S, there is probably an acidic gas present that is causing the reaction and resulting in those sensors to produce the registered values. Most of the interfering gases themselves are toxic, so when readings are found on the detection devices, fire fighters should be aware that there are toxic gases present. Some of these gases are also flammable, which means you may also get a reading in your flammable gas sensor. There are ways to determine which gases are present using detection devices a typical hazardous materials team carries.

- Electrochemical sensors can be easily overwhelmed and will max out regarding their readings. For example, most HCN sensors will top out at 50 ppm and will not tell you if you are in levels higher than the maximum. CO sensors usually have a maximum reading of 500 ppm. High exposures to a gas that causes a reaction will result in the sensor failing sooner than its intended life.

- Electrochemical sensors fail to the "0" (zero) point. With O_2 that's not necessarily a problem, since 0 percent oxygen is dangerous, and

you would not enter a potentially hazardous environment. But when a CO or HCN fails and reads 0, it may not indicate it has failed, and that can set up responders for a dangerous situation where they believe a toxic gas is not present.

- Some detection instruments will indicate that a sensor has failed, while others may not. Responders should ensure their devices are calibrated according to the manufacturers' recommendations and at a minimum should be bump tested prior to use. Many manufacturers recommend that the devices be calibrated before use, which is not always practical in an emergency response environment, but it should be bump tested. To bump test a device, it is exposed to a known quantity of gas to see if the meter responds appropriately. It may not read exactly to the levels of the target gas, but it should respond reasonably close to the level indicated in the bump test bottle. You can use the calibration gas to bump test your instrument, and it only takes a few seconds once a device is warmed up.

- Responders must be mindful of the reaction time of electrochemical sensors as they can take 20–30 seconds to react to an environment and may take even longer to clear back down. Some HCN sensors will take almost 3 minutes to react.

Another type of sensor that may be used by first responders is a photoionization (PID) sensor. The PID sensor is describe in more detail in Chapter 7. A PID in a smoke situation will detect the toxic soup that is in the air and tell you that something is present, but it will not tell you what exact toxic gas is present in the air. Responders should be aware of differences in the reaction time of the various detection devices: Electrochemical sensors react in 20–30 seconds (or more), LEL sensors in 7–10 seconds, and the PID reacts within 1–2 seconds.

Another option for the detection of toxic gases is colorimetric sampling using colorimetric tubes. Colorimetric tubes can also be used to confirm the readings of the electrochemical sensors. The colorimetric tubes are designed to determine the levels of a known gas, but they can also be used to help determine what unidentified gases may be present. Colorimetric tubes are a challenge as they will only identify the atmosphere in the immediate area of the tube being used and, in most cases, can only be used once. It can be a challenge to check a large structure for hazards. Furthermore, preparing the tubes for use may be difficult on the fire scene when scene lighting is limited.

There is one other device that may be of assistance for responders to detect toxic fire gases. The

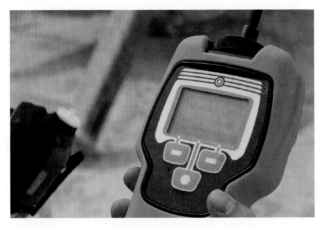

FIGURE 13-5 The ChemPro 100i has an overhaul library that can detect the common fire gases.
Courtesy of Christopher Hawley.

ChemPro 100i detection device, which is described in Chapter 11, has a detection library dedicated to fire gases (**FIGURE 13-5**). The libraries are set up to alarm to various predetermined action levels for the chemicals in the library and in the overhaul library; the device will alert you to the presence of common fire gases at levels above the permissible exposure levels (PEL). The device will not tell the user which gas has caused the alarm, but it would be one in the overhaul library—toxic soup. More advanced users can switch to one of the TIC libraries, which can aid in a presumptive identification of the specific fire gas.

A potential concern about the use of a ChemPro 100i is the cost of ownership. The initial outlay is high, but the sensors have a 5-year life or are at least guaranteed for the 5 years. Compare this with conventional electrochemical sensors that have a 1- to 2-year life and require frequent calibration. (Consider also that the calibration gas has a 6- to 12-month shelf life and can be expensive for the exotic gases like ammonia and HCN gas.) Responders could potentially use WMD grant funds to make the initial device purchase, as the primary design is for the detection of WMD, and it can be used in that capacity by the agency/department. The device can also be used for a variety of emergency response incidents and would be a useful and versatile tool in the detection device toolbox. At a minimum, a department's hazardous materials team should have this device, which could be called in for fire responses for the purposes of detecting fire gases. The hazardous materials team should also have a colorimetric tube set and a host of other detection devices, and they should be well trained in monitoring and detection and typically available for response. One of the other devices that can be used is the gas chromatograph/mass spectrometer (GC/MS), which can provide the best determination as to what is present. The use of a GC/MS is slightly problematic

as it would take some time to take multiple samples, but this device would provide definitive answers as to what is present.

Although we have already discussed some of the issues with the detection devices described in this chapter, they currently are the best tools at the first responder level, even with their respective challenges. Responders should develop a strategy for the use of detection devices and understand the limitations of any tool or device. The use of detection devices during active firefighting is not a good idea, or at least in the dense smoke. The heat and water can be immediately or cumulatively detrimental to these instruments. Many are water resistant, but most are not waterproof and can be damaged in temperatures above 100°F (38°C). Monitoring should be conducted when active firefighting has ceased and there is a possibility that personnel may remove their facemasks. The important factor is that monitoring should be conducted *prior* to any personnel removing their masks. Monitoring can, however, be established early on in the incident in locations away from the fire such as staging, the pump panel, the command post, rehab, etc.

Where to sample for the toxic gases can also present a challenge for responders, and it is largely dictated by the size of and the configuration of the fire building. The best strategy for the protection of the responders is to have monitoring occur at any location where personnel are operating, but monitoring should be done throughout the building to make sure no "toxic hot spots" are present. The toxic twins will be present, but in most cases, CO will be found throughout the building and may linger. HCN will not be as prevalent at high levels in most cases, but there will be areas where HCN may have accumulated. Any room or area that has a high concentration of plastic may have more HCN. Some forms of insulation will off-gas HCN, so it is possible that the HCN could be throughout the building. Both CO and HCN are both lighter than air in normal conditions, so they will be found in higher concentrations on upper floors. Even in a basement fire scenario, personnel could be exposed to high amounts of CO and HCN while checking the basement space. Situations where there is little actual fire but large quantities of smoke are present could also mean that CO and HCN are present. It is worth repeating that standard electrochemical sensors have a delayed response of 20–30 seconds, so when checking for toxic gases, it is important that responders using the devices move slowly through the structure or area. The level (4–6 feet or 1.2–1.8 m) at which people will be breathing the air in a space is the best place to do the monitoring.

Even though the toxic twins are lighter than air, when going below the fire area, monitors should always be used. Because the heat causes the smoke and gases to be pressurized, the products of combustion will be pushed in every direction. There are only 11 toxic gases that are lighter than air; the remaining gases and vapors (and there are lots of them) are heavier than air. Table 13-1 provides the density of some of the fire gases and where they will be found in relationship to the air.

It is important for responders to monitor continuously as conditions on an incident can change. During overhaul, crews may identify the presence of a toxic gas at a relatively low level initially, but during normal events on the scene, they may uncover hot smoldering debris, which could then produce higher levels of CO, HCN, or other toxic gases.

Operating in well-ventilated areas will obviously reduce the risk to the fire fighters. The use of positive pressure ventilation will increase the air flow, thus improving the air quality. Responders must exercise caution with the use of gas-powered positive pressure ventilation (PPV) fans in situations where there are low levels of CO, as the gas-powered PPV fan may add to or even increase the CO levels. In fire situations, this isn't a concern as the fire smoke will have higher levels of CO, but in situations where gas-powered PPV is used and there are low levels of CO or other gases initially, the CO level will increase depending on how well the PPV engine is operating. A fan that is not operating well can introduce CO levels above the REL in a building that previously did not have any CO. It is helpful to assess the gas-powered fans ahead of time to understand their operating levels.

Making Sense of Detection Findings

What can responders make of all this information? The bottom line question most of you have is probably this: Once I identify the presence of a toxic gas, at what level can responders remove breathing apparatus masks? The easy answer, and the one that affords the highest level of safety for responders, is that since responders are dealing with a toxic mess, you can take your mask off when you are getting ready to leave. This stance offers the highest level of protection from acute and chronic health hazards. Realistically, on incident scenes this may not happen; if that is the case, the detection methods outlined in this chapter can be used along with the NIOSH recommendations to make decisions about when to doff breathing apparatus masks. **TABLE 13-3** lists the recommended action levels at which SCBA should be used for the specific gases listed. Remember just measuring for CO and finding low levels does not mean there is not a large quantity of HCN or another gas present. Measuring for one or two gases can offer a

TABLE 13-3 Recommended SCBA Use Levels

Gas	SCBA Recommended for Levels Above
Carbon monoxide (CO)	35 ppm
Hydrogen cyanide (HCN)	5 ppm
Hydrogen sulfide (H$_2$S)	10 ppm (10 minutes)
Oxygen (O$_2$)	<20.8% or <21%

© Jones & Bartlett Learning

false sense of security, as there are many potential gases, not to mention the potential harmful particulates found in fires that are being inhaled and ingested.

There are many fire departments that set detection devices to alarm at lower levels for CO, such as 20–25 ppm, which increases the safety of their personnel. Even at those levels, it would not be uncommon to have a headache after an exposure if the exposure is more than a few minutes. Other departments state unless the readings are "0", then SCBA must be worn.

When you look at the tasks being performed when masks are commonly removed they are typically being done in an environment with an increased temperature, the fire fighters' core temperatures are higher, the pores are open, and pulse and respirations are increased. Each of these factors enables toxins to enter the body at a faster rate.

Fire smoke is a serious concern, and it is one component of what causes premature deaths of fire fighters. It can not only be dangerous and deadly at the time of an incident, but accumulations of smoke exposures over the course of a responder's service can be equally as dangerous and deadly over a longer period of time. In other words, it may not always be an acute exposure, but chronic exposure can kill you if you do not wear your SCBA.

Technology is getting better so that we are better prepared to detect the hazards we are facing on the fire ground, but it is not perfect. Most of these toxic gases cannot be seen or smelled, so even if you cannot see the smoke, it does not mean there is not a toxic soup waiting for you. Only the detection devices can tell you if the intended target gas is present, and you will need to monitor for the variety of toxic fire gases to ensure the safety of the responders present. Sampling for just one gas is a start, but it may not be enough as there are many toxic gases present that can cause harm.

After-Action REVIEW

IN SUMMARY

Fires are a toxic soup of hazardous gases and present risk to responders who do not wear respiratory protection. The detection of these dangerous gases is also quite a challenge due to the extreme conditions and the large number of potential materials present. Hazardous materials response teams have a number of detection devices that can be useful in determining some of the risks posed by fire situations.

KEY WORDS

Fire smoke The products of combustion, including gases, that produce a toxic atmosphere for anyone exposed.

Toxic twins Carbon monoxide and hydrogen cyanide, which are commonly found in fire smoke and have a synergistic effect on the body.

Review Questions

1. What are the two most common gases that responders look for in fire situations?

2. What are some challenges to using electrochemical sensors?

3. What are some challenges when using HCN sensors to detect HCN?

CHAPTER 14

Response to Sick Buildings

LEARNING OBJECTIVES

Upon completion of this chapter, you should be able to:

- Describe the sick building categories
- Describe the causes of sick buildings
- Identify acute sick buildings
- Describe response actions

Hazardous Materials Alarm

The hazardous materials team responded to a high-rise building for a chemical odor. Several people were affected with headaches and other minor symptoms. The occupancy was an office, and there were no unusual chemicals within the building. The team went through the sick building protocol listed later in this chapter and found that there was some form of alcohol in the air. The team searched throughout the building, as there was alcohol through the whole building. When the team was not able to locate the source of the alcohol in the building, they decided to check outside the building. They found there was alcohol in the air, which would be considered unusual. A flavoring production facility a few blocks away and upwind was found to be releasing alcohol vapor, which was brought into the fresh air intake of the building. When looking for a source of a problem, look inside and outside the building.

1. What levels of chemical contaminants would you anticipate finding in a sick building?

2. If there are symptomatic people in the building, is this a chemical or a biological problem?

3. What is your detection protocol for a response of this type?

Introduction

One response that hazardous materials teams respond to is unusual odors in a building, which would be put into the category of a **sick building**. Although this categorization is not quite technically accurate, it is the best way to describe the situation. The hazardous materials response to a sick building is unique, but this type of response is well within the capabilities of a well-trained, educated, and equipped hazardous materials team (**FIGURE 14-1**). Another factor to consider is a terrorist attack, in which the terrorists attack a building with weapons of mass destruction (WMD) material. Using the recommendations in this chapter, the hazardous materials team can determine if the problem is accidental or intentional or could be a terrorist attack. The key to solving a sick building problem is to first think outside the normal response box. Use ordinary hazardous materials tools in a unique fashion and follow some guidance but mostly follow your instincts. The cost for a building owner to hire an air quality contractor is extremely expensive. The actual cost to a hazardous materials team is minor, other than personnel costs. A hazardous materials team that is diversified in its variety of responses is one that will easily survive and, in many cases, thrive in a budgetary crisis.

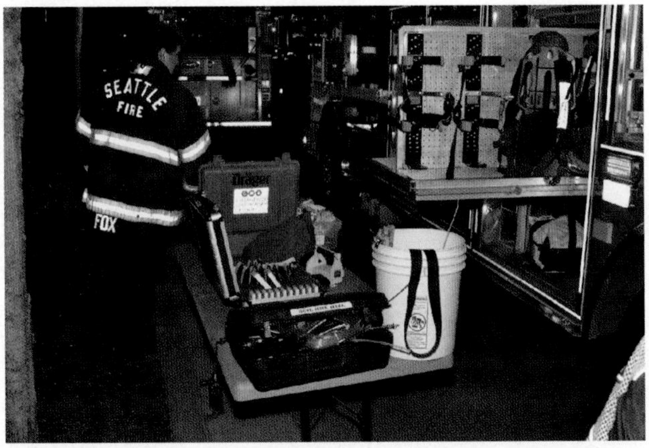

FIGURE 14-1 A well-equipped and trained hazardous materials team can conduct effective investigations at sick buildings.
Courtesy of Christopher Hawley.

> **LISTEN UP!**
>
> The true definition of a sick building is a building that causes the occupants to be affected by an unknown source.

A hazardous materials team can make a difference in a sick building because in many cases the problems found in buildings do not fit the exact definition of a sick building. This chapter focuses on sick building causes that can be detected or identified by a hazardous materials team.

The field of indoor air quality is very broad, and this chapter focuses on the most common issues. The true definition of a sick building is a building that causes the occupants to be affected by an unknown source. The health effects can vary from minor irritation of the respiratory system to acute life-threatening illnesses.

ASPECTS OF *SICK BUILDING* SYNDROME VERSUS BUILDING-RELATED ILLNESS

Sometimes the medical condition that affects the occupants of a sick building is called **sick building syndrome (SBS)** or **building-related illness (BRI)**. In either case, the Environmental Protection Agency (EPA) and the World Health Organization (WHO) have established some criteria to describe both medical conditions. A person who has SBS has symptoms of acute discomfort, with headaches; irritated eyes, nose, and throat; dry cough; itchy skin; dizziness; nausea; difficulty in concentrating; fatigue; and sensitivity to odors. The symptoms cannot be tied to a specific cause or event. The key identifier of someone with SBS is the fact that the symptoms disappear after he or she exits the building.

Someone with BRI may have a cough, chest discomfort and tightness, fever, chills, and muscle aches. The symptoms have identifiable causes and do not clear after exiting the building. Another term that is becoming associated with sick buildings is **multiple chemical sensitivity (MCS)**, in which it is thought that a person who has worked in a sick building develops severe reactions to many different chemicals. Even exposure to food products, cologne, or chemicals can cause severe reactions in some people. There are multiple issues with each of these described health effects, and it is best to try to avoid labeling a building until all the facts are determined.

FIGURE 14-2 Environmentally tight buildings are one of several reasons for the increased number of sick building responses.
Courtesy of Christopher Hawley.

The source of a sick building is usually the building itself or the contents. The number of sick buildings has increased in recent times and can be related to several causes, including building types and an informed public. New buildings are increasingly airtight or sealed buildings (**FIGURE 14-2**). With the increased concern for energy conservation, there has been an increase in sick buildings. The main problem is that the buildings do not allow for a good exchange of indoor air with the outside air, allowing a buildup of contaminants. Any small spills that in the past went out the ventilation system or were diluted by outside air coming into the building are now held in the building, causing problems. Problems with sick buildings have come to light because of an informed public and the advent of shock media. One person having a true issue with a chemical exposure in a building can be spread by the media as being a problem to everyone in the building. The media may use environmental issues as a platform to hype a small potential hazard into one that can cause problems to a larger group. There are people who are susceptible to chemical exposures, while others are not affected at all. The reaction to a bee sting is a chemical reaction and is one that we can use to discuss the body's response to chemicals. Assemble a group of about 50 people, and within that population group, there will probably be one to three people who are allergic to bee stings. They have

an immediate chemical reaction to the bee venom, while the remaining 47–49 people have no apparent reaction to the bee venom. Just as bee venom affects some people and not others, sick building chemicals have the same type of effect. If a population is exposed to a certain amount of a chemical, some will be immediately affected, while others will not show any effect. Isolation of the persons who are affected by a sick building chemical is an important issue, as psychogenic illnesses may start to develop. If you keep the people affected by the bee venom in the same group with those who do not have a problem with the venom, you will find that the persons who normally are not affected by bee venom will become symptomatic after a time. This sympathetic response is a common problem with sick buildings and is difficult to sort out.

The hazardous materials sick building varies from the previously given definition but is related to a number of causes, some of which are listed in **TABLE 14-1**. The hazardous materials sick building is usually related to something the occupants have caused and is a direct result of some type of chemical activity. The hazardous materials sick building is usually an acute issue as opposed to a chronic issue that a true sick building ends up being. Most of the causes of a hazardous materials sick building are easy to comprehend, and a direct factor can be determined. The building can provide any number

TABLE 14-1 Hazardous Materials Sick Building Causes	
Building	Outside factors
Furnishings	Neighboring facilities
Occupants	Friday 3:00 P.M.
Processes	

© Jones & Bartlett Learning

TABLE 14-2 Volatile Organics Detected in Carpets	
1,1, 1-Trichloroethane	Ethyl methylbenzenes
1,2-Dichloroethane	Hexanes
1,4-Dioxane	Hexene
4-Phenylcyclohexene	Methylcyclopentane
Acetaldehyde	Methylcyclopentanol
Acetone	Methylene chloride
Benzaldehyde	Octanal
Benzene	Pentanal
Butyl benzyl phthalate	Phenol
Carbon disulfide	Styrene
Chlorobenzene	Tetrachloroethene
Chloroform	Toluene
Dimethylheptane	Trichloroethene
Ethanol	Trimethylbenzenes
Ethyl acetate	Undecanes
Ethylbenzene	Xylenes

From Bayer C.W. 1990. Papanicolopoulos CD. Exposure assessments to volatile organic compound emissions from textile products. Indoor air '90. *Proceedings of the 5th International Conference on Indoor Air Quality and Climate*, Vol. 3, pp. 725–730.

of potential sources of illness, but it is usually related to the heating, ventilation, and air-conditioning (HVAC) system, which is discussed later in this chapter in the section Ventilation Systems. The furnishings may also be a cause in a sick building case, as new furniture will off-gas several chemicals and, depending on the air exchange rate, may cause a problem with some employees. Carpet is included under the furnishings category and, in many cases, is the culprit in sick buildings. Between the adhesive used to lay the carpet and the odor that the carpet itself presents, new carpet can be targeted as a cause of sick buildings. The common gases found in carpets are provided in **TABLE 14-2**. It takes about a year for new carpet to off-gas to a point that it no longer is an issue. The best time to install new carpet in offices or other commercial buildings is on a weekend, with lots of ventilation, to reduce the irritation to the occupants. The occupants may bring chemicals, usually mace or pepper spray into the building, causing problems. Also, the occupants may inadvertently mix some cleaning chemicals, which could cause problems. Even without mixing some cleaning materials, just their use may cause problems.

The industrial process is another issue that frequently arises with sick buildings. In many cases, these problems do not manifest themselves on the production floor but are found in the office area. The personnel working in the office are not used to the chemicals, as opposed to the personnel who work up close with the chemicals. If a new chemical is introduced to a process, it may cause problems at a facility, as personnel are not used to the odor.

One of the other possible causes are battery banks, which may be employed as a power backup system for computers or other electronic systems (**FIGURE 14-3**). When these batteries malfunction or overheat, they will produce hydrogen gas and, in some cases, sulfuric acid fumes. The hydrogen gas will not have an odor, but the sulfuric acid will present with a slight sulfur odor. Some incidents are caused by the batteries in exit signs.

LISTEN UP!

A NIOSH study in which several hundred buildings were studied for sick building issues found the following:
  53 percent had inadequate ventilation.
  17 percent had chemical issues from within the building.
  11 percent had chemical issues from outside the building.
  5 percent had biological issues.
  3 percent had contaminated fabrics in the building.
  12 percent were from unknown causes.

If there is a problem with the HVAC system, there is a potential for fume migration into the office area, causing problems. In a strip store setup, there may be a chemical process within the strip of businesses, but an occupancy three doors down is reporting the problem.

FIGURE 14-3 Battery bank, which can release hydrogen.
© Jones & Bartlett Learning

Several potential causes for this type of problem are the following:

- There may have been a change in the process, resulting in more chemicals in the atmosphere.
- The HVAC system may be malfunctioning, allowing migration into other occupancies.
- There may be a common attic allowing migration into other occupancies.
- There may not be good business-to-business wall separation, causing migration.

Outside factors and neighboring facilities can actually be combined but can be related to a number of issues. The predominant one is chemical activity that is occurring outside the building. Another building may be getting a new roof, and the tar odor can migrate to other locations. Some chemical processes involve a release of odors, which under normal circumstances is not an issue but occasionally presents problems. A weather inversion, which holds things low to the ground, may allow migration of odors into adjacent buildings. A malfunction or a change in the HVAC system may draw odors into adjacent buildings, causing problems. One other important outside factor involves **allergens**, which are discussed later in this chapter in the section Allergens.

The Friday at 3:00 P.M. consideration plays an important part in hazardous materials sick buildings. In many areas of the country, we can look to the first warm day of spring as a potential response day. Much like someone pulling a fire alarm to get out of work or school, the report of a

sick building usually results in the building being evacuated and may result in the employees being sent home. With the news media hyping sick buildings syndromes, these issues have become a big deal. Without an effective response and with an unidentified cause, the building owner or business manager may send the employees home for an early weekend. Labor and management are another issue that fits into this category. Separately ask both groups in the most diplomatic fashion about the status of the labor–management relationship, which may be a factor in why the sick building was reported and may help identify the cause.

> **LISTEN UP!**
>
> For the remainder of the text, we use the term "sick building" in place of "hazardous materials sick building".

Sick Building Categories

The sick building categories include acute, chronic on a short-term basis, and chronic on a long-term basis. The ability to recognize whether the problem is acute or chronic helps identify the problem. An acute sick building problem can be life threatening and usually does not meet the definition of a true sick building. A hazardous materials team can identify the cause of an acute sick building, as it is usually related to a chemical release within the building. Terrorism can cause an acute sick building type and can be used to panic the occupants of the building.

There are two types of chronic sick buildings: Short term and long term. Short-term hazards are usually chemically related and can also usually be identified by a hazardous materials team. The aspect of short-term time is related to hours and days. The true sick building cause can be difficult to identify, is long term, and relates to time in terms of weeks, months, and years. The causes of a long-term type of sick building are usually related to allergens or biological issues. The hazardous materials team may not be able to identify the exact cause but can usually point a building manager in the correct direction to make that identification. Some example time frames are shown in **TABLE 14-3**.

Causes of Sick Buildings

Sick buildings (SB) are basically the result of problems with the HVAC system, environmental factors, or contamination.

TABLE 14-3 Sick Building Time Frames

Sick Building Time Frames

The time frame associated with a sick building is a clue as to the possible source. The assignment of a time frame establishes a starting point for the investigation.

Time Aspect	Time Measurement	Common Causes	Usual Source	Assistance Required
Acute	Minutes	Chemical, mace/pepper spray, terrorism	Spill or other nonroutine release. Cleaning chemicals, construction work. Chemical attack	Usually none; hazardous materials team can handle
Acute	Hours	Chemical	Cleaning or construction, such as roofing, painting, carpet installation	Usually none; hazardous materials team can handle. Low levels may require long-term sampling.
Chronic	Days	Chemical	Construction, renovations, or allergen or cleaning. HVAC system	Indoor air quality specialist is usually required for a follow-up
Chronic	Months	Allergen or biological	HVAC system, building components	Indoor air quality specialist is required

© Jones & Bartlett Learning

Ventilation Systems

Identifying the building type is an important consideration in identifying an SB problem. The HVAC system varies with the types of buildings, and the type of HVAC system is a key issue with SB identification. The various types of buildings include residential (single- and multiunit), commercial, and high-rise. Although this is a simplification of the many types of buildings, the reason for this small grouping is the HVAC systems. The systems used in these three groupings are different from one another and help identify potential problems. The three systems vary by one basic principle, which is the amount of outside air that the units bring in. In general, residential heating and cooling systems do not bring in outside air. The units heat and cool the inside air and recycle it within the building. With the heating system, units that use some form of flame to heat bring in some outside air to allow combustion to occur. There is some outside air being brought in by all systems, but even at the maximum amount, it is not of a quantity to be of any concern. The fresh air exchange within a home relies on fresh air coming in from windows and doors. In this age of environmentally sound buildings, this can be problematic if there is limited movement through the doors. There can be the spread of a contaminant throughout a building using a residential HVAC system, but the source of the problem exists within the building. If someone intentionally spreads a material in a home, it would generally be placed within the HVAC system, either before or after the blower system. Placing it before the blower would mean it would have to get through the filtration system. These filters can be removed, which can be confirmed by visual inspection.

Commercial HVAC systems are divided into several categories but for the most part mimic the residential system. Depending on the building size and makeup, the HVAC system is tailored for the building type. In the case of a single occupancy, the likelihood is that there will be only one system, and it will be much like a residential system, just larger. There will be limited fresh air brought in, although a more sophisticated system allows for a maximum amount of 20 percent of fresh air to be brought in, although this amount should be considered extreme. Other commercial occupancies, such as a strip mall, generally have multiple HVAC systems, one or more for each occupancy. There is a possibility that odors from one occupancy can be picked up by another tenant's HVAC system, although this would be a rare occurrence as these systems are generally set up to bring in a minimum of fresh air.

High-rise buildings and larger commercial occupancies have systems designed to bring in fresh air to the maximum of 20 percent, but the amount is usually much less than that. It is possible for these systems to bring in a higher amount of fresh air up to 100 percent, but the normal settings are less than 20 percent. These buildings and any other HVAC system do not bring in a lot of fresh air because of cost. There is a cost to heat or cool air, and when the HVAC system is running, it is much cheaper to cool the inside air, which should be near 70°F (21°C), as opposed to cooling the outside air, which may be 90°F (32°C). The HVAC standards set by the American Society of Heating, Refrigeration, and Air Conditioning Engineers (ASHRAE) require that outside air be brought in and that a certain amount of air exchange take place. ASHRAE recommends that an air exchange of 20 cfm/person be done in an office, and 15 cfm/person in a residence. Areas such as smoking lounges require higher exchanges. These commercial systems vary the amount of outside air coming in by computer control to maximize cost savings to the building owner. One downside from the terrorism perspective is that if one is knowledgeable about the system, it could be used to spread material throughout a building. It would be challenging to attack a building in this fashion, but it is not outside the realm of possibilities. If a terrorist overrides the system to bring in 100 percent outside air, this change in air quality would be quickly detected, alerting the occupants to a problem in the building. They would not necessarily recognize terrorism, but it would immediately alert them that there was a problem with the HVAC system. Indoor air quality is discussed later in this section, but the key is that people in the building know when there is a problem and can identify the source.

Understanding how the system is set up is key to identifying the potential source of an SB event. **FIGURE 14-4**

provides a diagram of a typical high-rise HVAC system and identifies the various parts of the system. There are several main parts of the system that are standard, building to building, HVAC system to system. These parts include fresh air intake, return air, mixing box, heating/cooling system, and the main fan. From the terrorism perspective, there are good places to place an agent for dispersion, and conversely, there are places that would not distribute the agent at all. Unless altered, the fresh air intake brings in a minimum amount of air, which is used to make an exchange with the return air from the building. When the temperature is near 70°F (21°C), the system will bring in more fresh air, as it does not require any heating or cooling. When the temperature varies from that point is when the percentage starts to drop. Even on a day with 70°F (21°C) temperatures, if the humidity is extremely high, the system will not bring in a lot of air, as it would have to dehumidify it. On a high-rise, the fresh air intake is typically located in one of two places: The street level or the roof. Some newer buildings, especially those that may be targets, will locate the fresh air intakes in protected areas. When trying to locate the source of an unknown odor or other problem in a building, the fresh air intake is a good place to start. One would think that this would be a good place to place a terrorism agent, but luckily, this thought process is flawed. First, the intake only brings in a minimum of fresh air, which is then diluted with massive amounts of return air within the building, which is all mixed. A building contains a significant amount of air, and the amount of contaminant required to create a hazard would need to be large. The size of the building needs to be considered, as does the extreme amount of dilution that would occur when a contaminant is placed into that environment. Another factor going against products moving up a fresh air intake is related to physics. The contaminant must be pulled to

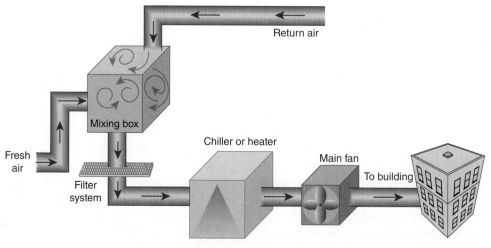

FIGURE 14-4 A typical high-rise HVAC system.

the top of the building, which, depending on the physical makeup of the chemical, may present some challenges. Some systems do have a great amount of pull, but keep in mind the height of the building and the size (and therefore the volume of air) of the air intake. Air intakes on the roof present an easy target for contaminants, as the system is pulling the contaminants down into the building, which is easy. We must consider dilution and the fact that the most it would be typically bringing in is 20 percent or less. For rooftop mounted intakes, there are also exhaust vents for various systems, which may be venting chemicals. In this scenario, these contaminants, can be drawn into the fresh air intake. Weather may be a factor in contaminants entering the system, as a weather condition that involves high humidity, fog, or an inversion will keep the contaminants low and hanging around the roof. The fresh air intake is certainly one of the first areas that should be examined if you are responding to a chemical odor or other situation that affects the HVAC system. It is not likely to be the source, but it is one of the quickest areas to visually check and rule out as a possible source.

The return air system is another location into which chemicals can be drawn and spread into other parts of the building. Placing an agent here is more likely than the fresh air intakes, but some protection mechanisms are still in place. All the return air is mixed together, which is a tremendous volume of air, which is also mixed with the fresh air. The same complications exist for a return air system, where a contaminant introduced on the first floor must travel to the top floor; all the while it is being mixed with other air, being diluted. The chemical and physical properties must be such to be conducive to move in that direction. The draw of the return air is not very active and will not "hold" on to contaminants very easily once introduced to the system.

Once the fresh air and the return air are mixed, they are then filtered. Any particulate that may have entered the system should be caught by the filters, if they are well maintained and installed correctly. During allergy season, an easy fix may be to change the filters, as they may have been clogged with the allergens, dust, and/or dirt. If you think a particulate has been introduced to the system, the first place to check is the filters. Most commercial (and residential) filters are designed to trap tiny-sized particles and will capture almost all contaminants that may have entered the system. There are two basic types of filters: Low efficiency and high efficiency. Low-density filters remove the least number of contaminants. The more efficient a filter is, the higher the cost. Another factor is the increased workload on the system the higher the filter efficiency. There are filter systems, actually a series of filters, which can protect a system against weapons of mass destruction (WMD) attack. These are expensive, require frequent changes, and are used for high-value target buildings. The type of filter used determines what the filter catches. Gases, vapors, and aerosols are not all caught by the same type of filter. Aerosols are the problem as they may be composed of solids or liquids suspended in air (**FIGURE 14-5**).

There are various filters based on the intended use or facility. Most homes use filters for dust, while a hospital or a clean room will have filtration systems that

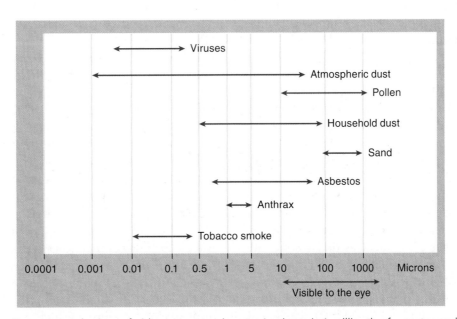

FIGURE 14-5 Some example sizes of airborne contaminants. A micron is 1 millionth of a meter and 1 inch equals 25,400 microns.

TABLE 14-4 Filter Contaminants

Use	Items Filtered	Particle Size (microns)	Filtration
Residential	Pollen, dust, dust mites	>10	<20%
Light industrial	Dust, mold, spores	3–10	35–70%
Industrial	Dust, Legionnaires	1–3	85–90%
Hospitals	Smoke, bacteria, fumes	0.3–1	90–95%
Clean rooms, surgical centers	Chemicals, biological, viruses	<0.3	99.97–99.9999%

© Jones & Bartlett Learning

are >99 percent efficient in removing many potential contaminants. The range of filters and contaminants caught are included in **TABLE 14-4**.

Once filtered, the air then enters the chiller (or heater), where it is cooled or heated. The chiller is a system of pipes that has cold Freon moving through it, causing the pipes to be very cold. Part of this process, though, involves the fact that water (condensate) will be running down the chillers, generally into a drip pan. This water is significant, as it will trap and hold any contaminants that may have gotten past the filter system. Any chemicals that are in the air will also be caught by this water system and held or altered. Chemicals that are easily broken down or are soluble in water will be caught by this portion of the system. The chillers may be part of the problem, as this is a common location for bacteria and other growth to be found. There should be antibacterial tablets in the drip pans to minimize the bacteria buildup. In some cases, HVAC personnel have placed HTH™ tablets in the drip pans, which cause large amounts of chlorine to be introduced into the system, presenting a great risk to the occupants of the building. These tablets are made for 10,000–20,000 gallons of water, not the 10 gallons that may be found in a drip pan. They make specific tablets for these systems, and they are the only chemicals that should be introduced in the drip pan. If you find bacteria or other growth in the chiller or in the drip pans, you may have found your problem. A good place to look for bacteria is on the discharge side of the chiller area, as this is not easily accessed and may not be cleaned regularly. A good HVAC system does not have any growth in the ventilation shafts or other parts of the duct. In winter, the coils can be 300–400°F (149–204° C), which will alter many chemicals as they enter the system. Even in the heating season, there should not be any growth

anywhere in the system, and if you find some, you may have solved the problem.

After the air is cooled or heated, it is then grabbed by the main circulation fan, which pushes the air to the building. In many systems, this is the only fan, and through tapering and shaft sizes, the amount of distributed air is modified. Some buildings have variable air flow volume boxes (helper fans) that allow individual floors or work areas to increase or decrease the amount of air flow to that area. Although rare, there have been cases in which these individual areas have been the source of the problem, and they should be examined for contamination. In many cases, these helper fans have belt problems, emitting a burning rubber-type odor. If the odor is isolated to a specific work area or a floor, then it is the helper fan. If the odor is building wide or on multiple floors, then the problem lies with the main fan. The best source of information about the buildings HVAC system is the mechanical engineer or the maintenance staff. As soon as you arrive at the event, you need to grab these people and hold on to them. They not only can answer questions about the system but will also be knowledgeable about other activities in the building that are likely candidates for the cause.

In one complicated response, the hazardous materials team was requested to respond to a high-rise building with chronic air quality issues (**FIGURE 14-6**). Although the building was known to have problems, this was the first time the hazardous materials team had been called to investigate the problem. A victim who was complaining of chemical irritation was on the sixth floor and had been removed by emergency medical services (EMS). She had climbed the stairwell to her office when she became symptomatic. The building consists of all offices, and within the high-rise component, there were no chemicals. The building was unusual in that the high-rise was surrounded by a one-story ring of

FIGURE 14-6 A high-rise building that once had sick building issues.

Courtesy of Christopher Hawley.

FIGURE 14-7 This high-rise is surrounded on the first floor by commercial occupancies, and the ventilation flowed from these areas into the high-rise.

© Jones & Bartlett Learning

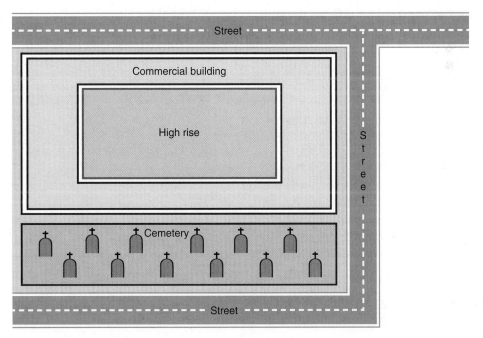

FIGURE 14-8 A graphical drawing of the setup of the high-rise.

© Jones & Bartlett Learning

commercial buildings (**FIGURES 14-7** and **14-8**). When the hazardous materials crews checked the lobby and the sixth floor, they did not locate any contaminants. They went to the penthouse to check the HVAC system and noted that there was a tremendous air flow coming up the stairwell toward the roof. To return to the lobby, they took the elevator, and when the elevator opened, there was a rush of air coming from the lobby heading up the elevator shaft toward the roof. Directly across from the stairwell was a hair salon, which was busy on this Friday evening, and there was a distinctive chemical odor coming from the salon. The air flow

FIGURE 14-9 After the hazardous materials team identified the air flow issue, the building owners added new HVAC systems to the one-story occupancies separate from the high-rise.
Courtesy of Christopher Hawley.

from within the high-rise was pulling contaminated air from the salon into the stairwell. The commercial buildings that surrounded the first floor did not have any mechanism to vent to the outside. The theory was that when the victim climbed the stairwell there was a chemical odor, and due to the physical exertion she was suffering some effect. Some psychological stress may have been included due to the problems with the building. The hazardous materials team suggested the building owner explore changing the air flow to vent the high-rise downward to flush the contaminants out from the first floor. A few months later, HVAC systems showed up on the first-floor commercial buildings (**FIGURE 14-9**).

Environmental Factors

Many of the issues related to response to a sick building are related to the quality of the indoor air. Two of the biggest factors involved in this quality are temperature and humidity level. Just a few degrees off and all occupants of the building will feel the effect.

> **LISTEN UP!**
> Two of the biggest factors involved in indoor air quality are the temperature and the humidity level.

Humidity has two extremes, which may be a problem, as too much humidity combined with an elevated temperature will cause many people to feel ill. The amount of carbon dioxide (CO_2) in the building is a clue as to how efficiently the HVAC system is functioning. Checking

the level of CO_2 inside the building as compared to the outside can be an indication of the air exchange. There are no hard-and-fast rules to what will be found in a building, as each building is different. There will always be some difference inside as compared to the outside, with the inside having more CO_2, generally 200–400 ppm more. A difference of more than 700 may indicate a problem. The buildup of CO_2 comes from the people inside the building. There are other conditions that would cause a buildup in a building, such as an inversion, but the inside and the outside will have comparable amounts of CO_2. It is important to make sure the CO_2 is coming from an inefficient HVAC system as opposed to a leak of CO_2 within the building. When you have elevated CO_2 in a building, people develop headaches, nausea, and general malaise (tiredness). Several exposure levels associated with carbon dioxide relate to sick building investigations. For business and industrial buildings, the threshold limit value (TLV) is 5000 ppm, and the short-term exposure limit (STEL) is 30,000 ppm. The World Health Organization (WHO) recommends that levels should be less than 6700 ppm, and the Canadian recommended exposure limit for homes is less than 3500 ppm. As for signs and symptoms, headaches are found when the levels are as high as 20,000 ppm. Acute irritation is found at levels exceeding 50,000 ppm. At levels in the 30,000-ppm range, there is some impairment noted over a period of days.

> **LISTEN UP!**
> The amount of carbon dioxide (CO_2) in the building is a clue as to how efficiently the HVAC system is functioning.

As was discussed in Chapter 4, there is a large amount of CO_2 stored in restaurants and bars (**FIGURE 14-10**). These tanks hold refrigerated liquid CO_2 for use by soda dispensing machines (**FIGURE 14-11**). Some restaurants will have a backup cylinder of gaseous CO_2 in case the larger tank runs out before it is refilled (**FIGURE 14-12**).

Contamination

The types of contaminants can vary within a building, including allergens, bacteria/viruses, and chemicals. Responders can identify as well as quantify some of these problems, while for other problems, the only task available to a responder is one of possible characterization. It is sometimes hard to identify a difference between the three large categories, but each has a distinct mechanism of harm and form of identification.

FIGURE 14-10 This fast food restaurant has a liquid CO$_2$ tank, and the tank to the left is the vapor holding tank.
Courtesy of Christopher Hawley.

FIGURE 14-12 This setup has a liquid tank and a gas cylinder backup.
Courtesy of Christopher Hawley.

FIGURE 14-11 This is the opposite end of the CO$_2$ system. The CO$_2$ lines may travel across the building.
Courtesy of Christopher Hawley.

Chemicals

Chemicals in a building are the easiest contaminants for responders to identify and quantify, but they are also difficult to deal with, as the amounts in the air are usually quite low. There are a variety of issues that arise when dealing with chemical problems in a building. Usually the occupants have done something that has caused the problem, so a simple investigation can usually help solve the problem. When the building

itself has chemical problems, the investigation becomes much more difficult.

The typical scenario for an occupant-caused problem is usually cleaning material or construction. The amount of fresh air being brought in will determine how much of the contaminant will be spread throughout the building. One of the major issues is that in this shock media–educated public the fact that the public assumes that because they can smell a chemical this smell equates to a direct harm to their body. The longer this believed harm occurs, the more you are unlikely to convince them otherwise. Many chemicals that have potential to cause harm have odor thresholds below that of the level where they begin to cause harm. Some people who are more sensitive to chemical exposures than others may incite mass illness among the other occupants of the building. It is important to separate out those individuals who are having problems from those who are not having any signs or symptoms. Leaving them together will invariably cause the nonsymptomatic folks to suddenly pick up the signs and symptoms of the other occupants.

One of the issues with painting, resurfacing, or sealing of floors, walls, and ceilings is that the odor is usually pretty irritating but may be below the detectable limits of most detection devices. The parts-per-billion (ppb) photoionization detector will pick up these odors, however, and is an essential tool for sick building response. A standard photoionization detector will not pick these levels up unless they are at higher amounts. When cleaning chemicals are involved, the amount of chemicals in the air is usually high enough to get readings on standard detection devices. While conducting monitoring survey, it is important to interview the maintenance staff to find out what they were doing and what the location was. In some cases, simple carpet cleaning is enough to irritate some occupants of the building. If bathroom-cleaning materials were inadvertently mixed, there is usually an off-gassing that occurs that will be irritating to some people. If this has occurred, the cleaning staff is usually not willing to admit to this inadvertent mixing, but through detection devices and good questioning techniques, the problem can usually be solved. When looking for cleaning chemical-type problems, it is important to search trash cans and dumpsters, as the culprits may have tried to hide their mistakes.

Any type of painting, resurfacing, or cleaning that will output irritating odors should be done when there are no occupants in the building or when there is an absolute minimum of occupants present. Obviously cost drives this issue, but when the building is closed and the complaints are being investigated, there is a tremendous cost to the occupant's business as compared to some overtime for the work crews. Many of these cases become workers' compensation issues and are tied up in court for many years, costing considerable amounts of money.

Other chemical problems that occur in buildings are caused by the building or its furnishings. According to the text Environmental Sampling for Unknowns, studies have found that new buildings can have up to 30 ppm of volatile organic compounds (VOCs) such as those found in Table 14-4. The predominant chemical usually found in these buildings is formaldehyde, along with several other VOCs. After a year, the level drops to less than 1 ppm for the same building. You should not find levels greater than 1 ppm for VOCs in office-type occupancies or other nonindustrial businesses. Many of these issues also apply when remodeling is done as well. New carpets, wall coverings, and floor tiles all release chemicals that have been known to cause irritation in some persons. A list of these materials is found in **TABLE 14-5**. The level of chemicals in the air may not be readily identifiable to us, but the occupants in the building are there for extended periods of time. New furniture such as desks, chairs, couches, and other office accommodations all release VOCs as well, some of which are chronic irritants. Office equipment such as copiers, printers, and other reproduction equipment also release irritating chemicals. In some businesses, there is an office area with an adjacent production area. Look to the production area for the source of the problem. There may be an area where chemicals are used but

TABLE 14-5 Common Building Chemical Contaminants	
1,1,1-Trichloroethane	Methyl ethyl ketone
Acetone	Toluene
Benzene	Aldehydes
Cellosolve	Deacon
Cyclohexane	Tetrachloroethylene
Freon 113	Tetrahydrofuran
Methyl isobutyl ketone	Hexane
Pentane	Limonene
Xylene	Aromatic hydrocarbons
Amines	
Formaldehyde	

© Jones & Bartlett Learning

that is separate from the area in which the problem is being reported. It is important to ask if there is any new chemical use or if there has been a change in quantity or procedures.

Allergens

Allergens are a common problem and, in some cases, may fit within the true definition of a sick building. Although a lot of attention is placed on the chemical part of sick buildings, allergens are the most likely culprit. The growth of fungus or the level of fungi in the building air is highly suspect as a leading cause of sick building syndrome. Unfortunately, it is the most difficult to detect, and a specialist must be brought in. In some cases, the allergen issue is a transient one and is related to an unusual spike in allergens in the air.

A dirty building that undertakes a cleaning project may also put an unusual number of allergens in the air. Two big areas that are usually problematic are carpet removal and ceiling tile removal. These projects should be done after hours when no occupants are in the building.

If the HVAC system is dirty and dusty, then it is spreading allergens throughout the building. The filters within the HVAC system are a clue to the cleanliness of the system, and if they are clogged or filled with dust and dirt, they are compromised, and allergens are being spread. Some duct work has a fiberglass lining, which if wet, or if it has been wet in the past, may be the host to mold or mildew, which are allergens. If you find evidence of these items, then they can be a suspected cause. There have been no studies linking surface mold and mildew with illness. The presence of mold and mildew indicates a problem but cannot be directly tied to occupant illness. Mold and mildew cause an odor, which can be irritating over time, and that causes health problems with some people. Mold and mildew anywhere have the potential to cause health problems throughout smaller buildings. In larger buildings, there is still potential to cause problems, but it will likely not be building wide. Mold and mildew will cause building-wide problems if they are in the HVAC system. There should not be any mold, mildew, or other growth anywhere in the HVAC system. The HVAC drip pan is a common area for mold and mildew. These drip pans become great breeding grounds for all types of bacterial growth. In addition, if there are problems in the HVAC systems, the drip pans are suitable locations to add deodorizers or other cleaning compounds. Mushrooms growing in the system are a big clue that there is a problem. The cleaning of the HVAC system is another area that may cause problems, as the dirt and debris will be spread throughout the building. This cleaning should be done after hours when no one can be affected by it. In some cases, the cleaning involves the use of chemicals, which themselves may cause irritation to some people.

SAFETY TIP

The presence of mold and mildew indicates a cleanliness problem and cannot be directly tied to occupant illness.

One good question to ask is if there are roof leaks or other water problems, as they are sources of mold and mildew. If there has been a roof leak that has reached a suspended ceiling, then mold and mildew may be growing on the top side of the ceiling tiles. Even if you cannot see a stain on the bottom side of the ceiling tile, it is a good idea to look at a representative number of ceiling tiles to check for contamination. Check the building for other water leaks, such as in basements, as they are another common area for mold and mildew.

Bacteria/Viruses

Viruses generally spread through a building by human contact and normal spread of colds and other viral illnesses. The most likely routes of spread are related to

poor house cleaning, contaminated food preparation areas, bathrooms, telephones, and close proximity to the sick people. In rare cases, these viruses can be spread throughout the building by the HVAC system. A common problem within buildings that occurs several times a year is attributed to a bacterium known as *Legionella pneumophila*, commonly known as Legionnaires' disease. Each year there are several incidents of Legionnaires' disease that can cause fatalities. In the past 15 years, the number of cases has quadrupled. Legionnaires' disease became famous when it was found to have caused approximately 29 deaths and affected 189 people at an American Legion convention at a Philadelphia hotel in 1976. The number of cases is about 5000 per year according to the CDC. A milder form of Legionnaires' disease, known as Pontiac fever, has less severe symptomology. Several incidents each year involve fatalities, and typically a couple of persons die at each event.

- In 2014–2015 in Flint, Michigan, 91 persons became ill, and there were 12 deaths.
- In 2016, there was an outbreak in the South Bronx (NYC) where 12 people died.
- There were 127 cases in a Toronto, Canada, nursing home in October 2005 and 21 deaths.
- In a Baltimore, Maryland, hospital in July 2005, there were five cases with one death.
- Although typically linked to occupants of a building between November 2003 and January 2004, in Pas de Calais, France, there were 86 cases in the surrounding community, including 18 fatalities, from *Legionella*. An industrial cooling tower and wastewater system dispersed the bacteria into the adjacent neighborhood. The dispersion released the bacteria 3.7 miles (6 km), which was highly unusual.

There are 34 known species and 50 subgroups of *Legionella*. *Legionella pneumophila* thrives in stagnant water. This water can come from a variety of sources such as aerosolized water in misters, hot water heaters, showers, hot tubs, and humidifiers. Any place water stands for a period of time such as in drip pans is subject to the growing of *Legionella*. There have been cases in which the sources were grocery store fruit and vegetable misters, fountains, and a hot tub display in a hardware warehouse. The hot tub display case presents an interesting view of how *Legionella* can be spread. Several persons became ill in a suburban community, and it was determined that they had all contracted Legionnaires' disease. There was no common thread with the affected persons, which normally points to a source. After more than 30 days, it was finally determined that all the affected people had been to a hardware store, which had a hot tub display. The tub in question had been sold and luckily had not yet been installed in the purchaser's home but was sitting in his garage. Testing showed that the filter system contained *Legionella*, and through contact with the water, each of the affected persons contracted Legionnaires' disease. In 1999, two people died and several more become ill from Legionnaires' disease in a Baltimore manufacturing facility. It was determined that a production line rinsing system contained *Legionella*.

SAFETY TIP

Each year there are several incidents of Legionnaires' disease that can cause fatalities.

There are two other biological concerns, both airborne pathogenic fungi. The most common one is *Aspergillus*, which causes aspergillosis, and the other is *Histoplasma capsulatum*, which causes histoplasmosis. Hospitals are at risk for aspergillus, which results from a fungus growth, of which there are several forms. Although *Aspergillus* can be found in many buildings, patients in a hospital are in a weakened condition and are more likely to be affected. *Histoplasma capsulatum* is of concern for building occupants and fire fighters. This fungus grows in bird droppings, from pigeons or chickens. After a sufficient time, the bird droppings dry and become dust-like. When this dust and dirt are disturbed and become airborne, persons without respiratory protection are at risk for histoplasmosis. This problem appears in both rural areas and inner cities.

LISTEN UP!

Any place water stands for a period of time is subject to the growing of *Legionella*.

Acute Sick Buildings

There are four basic causes of acute sick buildings. The four are related to cleaning chemicals, mace or pepper spray, HVAC systems, and building chemicals. Most of the incidents will fit into one or more of these causes. The flow chart provided in Appendix A provides a breakdown of how to proceed for these types of events. By following the chart and using the checklist provided in Appendix B, you can solve most sick building issues. One item to always remember is that even though you may not be able to identify a specific cause, you will be able to identify what is not the cause.

The cleaning chemicals cause focuses on chlorine, ammonia, and acid gases, as they are commonly used or released when cleaning chemicals are the source of the problem. Obviously, storage areas are the focus of the investigation, as is any area that was being cleaned or had been cleaned. The reported odor by the occupants can help guide this type of investigation. If they report a petroleum-type odor, then colorimetric tubes for hydrocarbons should be used as well. The readings on a PID are crucial as well when these types of odors are reported.

Mace and pepper spray are a difficult issue to solve, as the chemicals are difficult to detect unless you have a ChemPro 100i or a GC/MS. You can identify mace and pepper spray using colorimetric tubes, but you must be in the building quickly, the building must have been tightly shut, and your response time must be quick. Mace can be detected (the hydrolysis product) by using the chloroformates (Dräger) tube, and pepper spray can be detected with the olefins (Dräger) colorimetric tube. Mace is easier to detect than pepper spray, and the pump strokes must be increased to make the tubes more sensitive. One way to determine the use of mace or pepper spray is to do swab tests on clothes or carpet and then use the GC/MS. Find the most affected persons and bag their clothes in a tightly sealed bag, then grab an air sample to run through a GC/MS. If you do not have access to a GC/MS, then you would have the samples taken to a local laboratory.

If the HVAC system is the cause, we can point the building owner in the right direction, but we cannot identify the specific cause. Temperature and humidity play a major factor in the comfort of the people in the building. Just by asking a few questions and then determining what the actual temperature and humidity are can sometimes solve a small problem. If the problem is the air exchange, then the problem is more difficult to solve. The gas to check in an air exchange system is CO_2, which should be slightly elevated inside a building as compared to the outside air. However, there should not be a large disparity between the inside and outside. More than a one-third difference in the values indicates a potential problem. If more than 700 ppm CO_2 is found inside a building, then the levels are high enough to cause some irritation with some people. Most buildings have CO_2 levels of 400–700 ppm and have a corresponding level outside of 200–500 ppm. The higher the amount inside means that there is not enough air exchange with the outside air. This issue is economic in some cases, as the more fresh air that is brought in requires additional heating and cooling. Just make sure the CO_2 levels are from buildup in the building, not a leaking CO_2 cylinder.

Other HVAC issues relate to the cleanliness of the system, which may indicate an allergen or biological hazard. It is outside the realm of a hazardous materials team to identify the exact type of allergen or biological that may be present, but we can point the building owner in the right direction. By identifying dirty filters, mixing box, chiller/heater, or duct work, a potential problem can be investigated. There should not be any dirt, dust, or growth within an HVAC system. If you find any, then refer the building owner to an HVAC cleaning specialist or indoor air quality contractor.

Related to the cleanliness of the HVAC system is the cleanliness of the building. Any accumulation of dirt or dust in the building, particularly in areas that may be picked up by the HVAC system, are always suspect for problems. Any moisture, mold, or mildew on the walls, ceiling, or floor is also an indication that there is moisture in the building.

The chemical issue is the easiest to determine, as responders' detection devices usually pick up a chemical. As detection devices such as the RAE ppb photoionization detector and the ChemPro 100i have detection limits extremely low, and in some cases below the odor threshold, the use of these devices can help solve a sick building mystery. Most chemical issues in a building are related to a human factor; someone in the building has done something to cause the problem.

Specific Response Actions

Once you have evacuated the building, shut down the HVAC system, and ensured that all the doors and windows are closed, you can start the investigation. Start interviewing the most injured or affected people, as they were most likely the closest person to the problem. While these initial interviews are taking place, the air monitors should be warming up, bump tested, and readied. The maintenance person and/or manager should be interviewed as well and told to stand by at the command post or hazardous materials unit. If you learn of chemicals stored in the building, or during the interviewing you discover that a chemical hazard may be the cause, determine the most effective detection method for those chemicals. Have the air monitoring crews start at the areas where the most affected people were located. If many people are affected, or the problem seems to be building wide or floor wide, then start your investigation at the HVAC unit, paying attention to the fresh air intakes and the return air duct. When dealing with a sick building, you should keep terrorism in the back of your mind. The probability is low, but many terrorism scenarios can be initiated through a sick building call. A response team that is effective in identifying sick building causes is a step ahead for terrorist attacks on a building.

If you can identify the cause of the sick building, then follow the appropriate strategy first, and follow up as needed with the other strategies. In the spring and summer, make Freon a high priority for sampling, and in winter, make CO a high priority. Always check for these gases out of season but move them to a lower priority in the off-season. Concentrate on the immediately dangerous gases first, and then work to the chronic hazards. When you have run through a sampling scenario, make the colorimetric tubes more sensitive, and do some more sampling. Have air monitoring crews meet with the personnel conducting interviews to compare notes. Have a crew member focus on the SB checklist to cover the other areas not covered in the first interviews. It is important to interview someone from management and someone from the labor group. The more people from each side the better, and make sure the interviews are conducted away from other people. It is important to document these interviews as well as to document your air monitoring procedures. The air monitoring use report is a good worksheet to record any readings and track the process. The worksheet has the sick building chemicals listed in the colorimetric section to make this sampling easier. This worksheet can then become part of the final report and documentation.

The air monitoring strategy focuses on two areas: The RBR profile to identify serious health risks and a focused strategy looking for specific chemical or biological hazards. The focused strategy is divided into four areas: Mace/pepper spray, cleaning chemicals, biological, and chemical. If you cannot identify which of these four areas to start with, follow them in order, as provided in the worksheet provided in Appendix B. No matter the situation, always take in pH paper, a three- or four-gas instrument, and a PID. The PID is most important to pick up tiny amounts of chemicals in the air and should be monitored most of the time. It is not likely that the LEL sensor will pick anything up in a sick building. A parts-per-billion PID is highly recommended. The ChemPro 100i device is also helpful with its trend analysis function. Oxygen is not likely to drop, but the CO is important to watch, especially during the initial entry. Any time you change occupancies, floor, or section, all monitors should be observed for changes. The initial monitoring should be done with the HVAC system shut down. After sampling the building, or if you suspect that the HVAC is the source of the problem, have monitors in place and then turn the system back on. Pay particular attention to the discharge vents and the return air vents. The sampling priorities for the focused strategies are outlined in **TABLE 14-6**.

When all the sampling is done, you have covered most hazards usually found in a sick building. If you

TABLE 14-6 Focused Sampling Strategies

Mace/Pepper Spray	Cleaning Chemicals	Unidentified Chemical	Allergens or Biologicals
Evacuate building.	Evacuate building.	Evacuate building.	Evacuate building.
Shut off HVAC system; shut doors, windows.	Shut off HVAC system; shut doors, windows.	Shut off HVAC system; shut doors, windows.	Shut off HVAC system; shut doors, windows.
Start interviews.	Start interviews.	Start interviews.	Start interviews.
Identify source location. Don appropriate PPE.	Identify source location. Don appropriate PPE.	Identify source location. Don appropriate PPE.	Identify source location. Don appropriate PPE.
Follow RBR sampling procedures: pH, PID, ¾ gas, PPB PID, FID, ChemPro, and FTIR.	Follow RBR sampling procedures: pH, PID, ¾ gas, PPB PID, FID, ChemPro, and FTIR.	Follow RBR sampling procedures: pH, PID, ¾ gas, PPB PID, FID, ChemPro, and FTIR.	Follow RBR sampling procedures: pH, PID, ¾ gas, PPB PID, FID, ChemPro, and FTIR.
Use GC/MS.	Colorimetric tubes: Polytest Chlorine Ammonia Formic acid Hydrochloric acid Ethyl acetate Benzene Acetone Alcohol Note: If no readings are observed, increase sensitivity.	Colorimetric tubes: Polytest Halogenated hydrocarbons Ethyl acetate Benzene Acetone Alcohol Phosgene Toluene Xylene Mercaptan Note: If no readings are observed, increase sensitivity.	Rule out chemical hazards. Use colorimetric tubes for: Polytest Halogenated hydrocarbons Ethyl acetate Formic acid Ammonia Note: If no readings are observed, increase sensitivity.
Colorimetric tubes: Polytest Chloroformates Olefins Note: If no readings are observed, increase sensitivity	Check HVAC system effectiveness with CO_2 and ozone colorimetric tubes.	Recommend further testing by indoor air quality specialists.	Visually examine the building for clues of dirt, dust, mold, mildew, or other moisture-related problems.
Check HVAC system effectiveness with CO_2 and ozone colorimetric tubes.	Recommend further testing by indoor air quality specialists.		Check HVAC system effectiveness with CO_2 and ozone colorimetric tubes.
Recommend further testing by indoor air quality specialists.			Recommend further testing by indoor air quality specialists.

did not find anything with the suggested strategies, you may want to sample with other available colorimetric tubes. If you sample and did not find anything, that does not mean that it was a false call. It means there are three possibilities: Either you did not have a detection device for the material present, or the level of contaminant is below the minimum detection limit for your detection devices. The last possibility is that the building was ventilated prior to the arrival of the detection devices. When you are done and did not find anything, you can be assured nothing present is an acute risk to the occupants or nothing present is immediately dangerous to their life or health. It does not mean that there is not a chronic or long-term hazard present. There are times when you know a material is present, but you are unable to detect it. In these cases, refer the building manager to an indoor air quality specialist. Some hazardous materials teams and some health departments can do long-term sampling. The air in these types of buildings needs to be sampled for a minimum of 8 hours, if not 40 hours, and the gas samples run through a gas chromatograph. If you suspect a biological or allergens, then specific testing needs to be done for these types of materials.

When referring, it is recommended that you know something about the companies that you are referring. Many hazardous materials teams maintain lists of hazardous waste contractors that they provide to a spiller so that a spill can be cleaned up. In most cases, the hazardous materials team knows the capabilities of the companies on this list and in unusual situations can make a recommendation as to which company may be

best suited for a particular spill. This type of familiarity is necessary for indoor air quality companies.

When conducting the investigation, it is best to keep all employees near the building so that you can ask questions. In some cases, there is a desire to send them home, but it is best to keep them around. Every so often have a representative brief them as to what is occurring and the time frame. If you did not find anything, do not tell the employees they are imagining things or that nothing is present. The standard line is "We tested for the following items . . . and we did not find any detectable levels. Based on these results, we did not find any acute or immediately life-threatening levels. We did not nor could we rule out long-term chronic issues, as that is beyond our capabilities." At this point, you should inform the employees of the mitigation plan. If you did not find anything, and you suspect that there really is nothing present (the Friday at 3 P.M. type of incident), then one option is to send the employees home for the day and have the HVAC system run with full 100 percent outside air for a period of time, such as overnight or over a weekend. This does have risks, as the employees may desire yet another day off and may dial 911 later to make that happen again. The flushing of the building gives the impression that you think something is there, but you just cannot identify it. Go with your instincts. Document what you did and why you did it. Be prepared to defend your actions in court 1–5 years later, so document very well. The reports, flow charts, and worksheets provided in the appendixes will assist with that effort.

After-Action REVIEW

IN SUMMARY

The response to a sick building can be very frustrating, but at the same time, it can be very rewarding, especially when you successfully identify a cause. When you solve a mystery, the employees are happy, the building owner is happy, and the hazardous materials team looks good to the community. A well-thought-out sick building investigation can aid in workers' compensation cases and assist in treatment of exposure victims; failure to provide effective monitoring for life safety risks may place the hazardous materials team in jeopardy. The actual cost is minimal, other than personnel, which should not be

considered anyway, as the customer is already paying that bill. The community outreach benefits not only the hazardous materials team but the citizens as well. The hazardous materials team gets some great practice in monitoring and the use of colorimetric sampling, with a minimum of stress. There is some initial work in the form of research that needs to be done so that you can become familiar with the indoor air quality specialists in your area. You may be able use them in other situations, as many have other expertise or specialized detection equipment that many hazardous materials teams do not.

KEY TERMS

Allergens Substances that cause adverse reactions, such as sneezing, in sensitive individuals.

Building-related illness Illness related to a building, the symptoms of which have identifiable causes but do not stop after the person leaves the building.

Multiple chemical sensitivity (MCS) A condition of severe reactions that is thought to result from exposure to various chemicals and other substances.

Sick building A building that is suspected of causing irritation or health problems for the occupants.

Sick building syndrome (SBS) A temporary condition affecting some building occupants with symptoms of acute discomfort that stop when the person leaves the building.

Review Questions

1. In terms of sick buildings, what is the main problem in new environmentally sealed buildings?

2. What is a common acute problem that hazardous materials responders face in sick buildings?

3. What percentage of fresh air does the fresh air intake usually bring in?

4. What are the four basic causes of acute sick buildings?

5. What is the key chemical that hazardous materials teams should determine the level of when dealing with a potential sick building?

6. During the initial investigation, should the HVAC be on or off? Why?

7. What is the most common form of fungus with the potential to kill the occupants of a building?

Glossary

Absorbed dose Measure of energy transferred to a material by radiation.

Accuracy Term to describe how closely a monitor's readings are to the actual amount of gas present.

Acid Material that has a pH of less than 7.

Activity Measure of the number of decays per second in a radioactive source.

Acute A quick, one-time exposure to chemicals.

Acute effect An acute exposure causes an immediate effect.

ALARA Radiation exposure reduction that means "as low as reasonably achievable."

Alkali Material with a pH of greater than 7. Also known as a base or a caustic.

Allergens Substances that cause adverse reactions, such as sneezing, in sensitive individuals.

Alpha Type of radioactive particle.

Amines Group of compounds with nitrogen (N) as part of its makeup.

AP2C French chemical agent meter (CAM).

Base Material with a pH greater than 7. Also known as an alkali or a caustic material.

Baumé scale An indication of the density of the acid.

Becquerel (Bq) Unit of measurement for the activity level of a radiation source.

Beta Type of radioactive particle.

BioCheck Protein screen for biological materials.

Biological indicating device A detection device for biological agents such as anthrax.

Biomimetic A type of CO sensor used in home detectors.

Boiling point Temperature at which liquid changes to gas. The closer the liquid is to the boiling point, the more vapors it produces.

BTEX Acronym for benzene, toluene, ethyl benzene, and xylene.

Building-related illness Illness related to a building, the symptoms of which have identifiable causes but do not stop after the person leaves the building.

Bump test Using a known quantity and type of gas to ensure that a monitor responds and alarms to the gases that are being tested.

C2 Canadian agent detection kit.

Calibration check Checking the response of a monitor against known quantities of a sample gas (most commonly against calibration gas).

Calibration Electronically adjusting the monitor to read the same as a calibration gas, which is an accurately known gas concentration.

Capillary pore sensor A sensor that uses a small tube to carry the gas into the sensor housing.

Carcinogen Material capable of causing cancer in humans.

Catalytic bead sensor The most common type of LEL sensor; uses two heated beads of metal to detect the presence of flammable gases. Also known as a pellistor.

Ceiling level Highest exposure a person can receive without suffering any ill effects; combined with the PEL, TLV, or REL, it establishes a maximum exposure.

Chameleon A colorimetric test kit for a variety of chemicals.

Chemical Abstracts Service (CAS) A service that registers chemical substances and issues them unique registration numbers, much like a Social Security number.

Chemical classifier A testing strip that includes five tests; a companion test strip is the wastewater strip.

ChemPro100i Warfare agent detection device.

Chip measurement system (CMS) A colorimetric sampling system that uses a bar-coded chip to measure known gases.

Chronic Continual or repeated exposure to a hazardous material over time.

Chronic effect Repeated exposure several times a day for a long period of time.

Civil Defense Simultest (CDS) Colorimetric warfare agent detection system.

Coliwasa A tube used to collect samples from a drum or tank; formal name is composite liquid waste sampler.

Colorimetric A form of detection that involves a color change when the detection device is exposed to a sample chemical.

Colorimetric tubes Glass tubes filled with a material that changes color when the intended gas passes through the material.

Combustible gas indicator A old name for a device designed to measure the relative flammability of gases and to determine the percent of the lower explosive limit. Also known as an LEL monitor.

Concentration The amount of a material in a certain volume.

Convulsant A chemical that can cause seizure-like activity.

Corona discharge (CD) A form of ionization that uses an electrical arc to ionize a gas sample.

Correction factor Factor that applies to how a given gas reacts with a monitor calibrated to a different gas.

Corrosive Material that can cause damage to skin or metal and is either acidic or basic.

Culturing Process of growing biological material.

Curie (Ci) Unit of measurement for the activity level of a radiation source.

Deoxyribonucleic acid (DNA) Substance from which genes are made.

Dosimeter Device that measure the body's dose of radiation.

E3500 Explosives detection device.

Electrochemical sensor A sensor that has a chemical gel substance that reacts to the intended gas and provides a reading on the monitor.

Electrolyte Material used to conduct electrical charges inside a gas sensor.

EST-4200 Ultimate Vapor Tracer Detection device that can detect drugs, explosives, and chemical compounds.

Etiological Includes biological agents, viruses, and other disease-causing materials.

EVD-3000 Explosive detection device.

False negative Detection device's failure to identify a target material when the material is present.

False positive Detection device's identification of a possible threat when one does not exist or incorrect identification of a contaminant.

Film badge Small piece of X-ray film in a plastic case with filters to identify radiation doses received.

Fire point The lowest temperature at which a liquid ignites and, with an outside ignition source, continues burning.

Fire smoke The products of combustion, including gases, that produce a toxic atmosphere for anyone exposed.

Fissile Descriptor of an isotope that can be induced to fission.

Flame ionization detector (FID) A device that uses a hydrogen flame to ionize a gas sample; used for the detection of organic materials.

Flammable gas indicator A gas monitor that is designed to measure the relative flammability of gases and to determine the percent of the lower explosive limit. Also known as LEL monitor, or combustible gas detector.

Flammable range The numeric range, between the lower explosive limit and the upper explosive limit, in which a vapor burns.

Flash point The minimum temperature of a liquid that produces sufficient vapors to form an ignitable mixture with air when an ignition source is present above the liquid.

Fourier transform infrared device Detection device that uses infrared light to detect solids, liquids, and gases.

Gamma Form of radioactive energy.

Gas chromatograph/mass spectrometer A detection device that separates mixtures and identifies them by their retention and travel times, then ionizes the separated chemicals and compares the ions to a library of known materials.

Geiger-Muller (GM) Detection device type for low-energy materials.

GID-3 A detection device for nerve and blister agents.

Gray (Gy) Measurement of radioactivity, equivalent to RAD.

Half-life Amount of time for half of a radioactive source to decay.

Handheld assay See handheld immunochromatographic assay.

Handheld immunochromatographic assay (HHA) Detection device much like a home pregnancy test for biological materials.

Hazardous Waste Operations and Emergency Response (HAZWOPER) An OSHA regulation that covers waste site operations and response to chemical emergencies. Found in 29 CFR 1910.120.

HazCat™ kit A chemical identification kit that can be used for solids, liquids, and gases.

Highly toxic Toxic material that can cause harm in very small doses.

Hydrolysis material The breakdown product(s) of a material in reaction with water.

Hyperbaric chamber A pressurized chamber that provides large amounts of oxygen to treat inhalation injuries, diving injuries, and other medical conditions.

Incapacitating concentration (ICt$_{50}$) Military term for incapacitating level set over time to 50 percent of the exposed population.

Industrial hygienist A person trained in occupational health and safety.

Inert A chemical that is not toxic but displaces oxygen.

Infectious Caused by pathogenic microorganisms, such as bacteria, viruses, parasites, or fungi; the diseases can be spread, directly or indirectly, from one person to another.

Infrared sensor A type of LEL sensor that uses infrared light to detect flammable gases.

Interfering gases Picked up by the sensor but not intended to be read by the sensor.

International units (IU) Unit of measure for pharmacists based on the effect on the human body.

Ionization potential (IP) The ability of a chemical to be ionized or have its electron removed. To be read by a PID, a chemical must have a lower IP than the lamp in the PID.

Ionizing radiation Radiation that has enough energy to break chemical bonds and create ions. Examples include X-rays, gamma radiation, and beta particles.

Ion mobility See Ion mobility spectrometry.

Ion mobility spectrometry (IMS) A detection technology that measures the travel time of ionized gases down a specific travel path.

Irritant Material that is irritating to humans but usually does not cause any long-term effects.

Laboratory Response Network (LRN) System of laboratories established by the CDC and the FBI for the identification of potential chemical or biological materials.

Lead wool sensors A sensor that uses lead wool in the detection of oxygen.

LEL meter The best name for a meter that is used to detect flammable gases.

LEL See Lower explosive limit.

LEL sensor A sensor designed to look for flammable gases; can be of four designs: Wheatstone bridge, catalytic bead, metal oxide, and infrared.

Lethal concentrations (LC$_{50}$) Term used to describe the amount of vapors or gas that was inhaled to cause harm or death to the exposed test animals. A lethal concentration of 50 percent means that the dose killed half the animals that were exposed to that level.

Lethal dose (LD$_{50}$) Term used to describe a dose of a solid or a liquid that caused harm or death to test animals. A lethal dose of 50 percent means that the dose provided killed half the animals.

Lower explosive limit (LEL) The lower limit of the flammable range; the lowest percentage of flammable gas and air mixture in which there can be a fire or explosion.

M8 paper A paper detection device like pH paper that detects liquid nerve, blister, and VX agent.

M9 paper A detection device used for liquid nerve and blister agents; comes in a tape form.

M256A1 kit A detection kit for nerve, blister, and choking agent vapors.

Mass spectrometer Almost always coupled with a GC, the MS ionizes chemicals and measures the weight of the given ions and compares it to a library of known materials.

Membrane sensor A sensor that relies on the gas moving through a membrane on top of the sensor.

Metal oxide sensor (MOS) A form of LEL sensor.

Microcantilever sensor Very small sensor with multiple coated levers that each react differently to groups of chemicals.

Microelectromechanical sensor (MEMS) Detection technology that uses cantilever sensors on a circuit board for the detection of a variety of materials.

Microrem (μR) Measurement of radiation; normal background radiation is usually in microrems.

Millirem (mR) Higher amount (1000 times) of radiation than microrem.

Molecular weight Weight of a molecule based on the periodic table or the weight of a compound when the atomic weights of the various components are combined.

Multiple chemical sensitivity (MCS) A condition of severe reactions that is thought to result from exposure to various chemicals and other substances.

National Fire Protection Association (NFPA) A consensus group that issues standards related to fire, hazardous materials, and other life safety concerns.

National Institute of Occupational Safety and Health (NIOSH) The research agency of OSHA that studies worker safety and health issues.

Neutron Form of ionizing radiation.

NFPA See National Fire Protection Association.

NIOSH See National Institute of Occupational Safety and Health.

Nonionizing radiation Radiation that does not have enough energy to create charged particles, such as radio waves, microwaves, infrared light, visible light, and ultraviolet light.

Nuclear detonation Device that detonates through nuclear fission; the explosive power is derived from a nuclear source.

Occupational Safety and Health Administration (OSHA) Government agency tasked with providing safety regulations for workers.

Organophosphate pesticide (OPP) Toxic group of chemicals that can cause seizure-like activities. Nerve agents are chemically similar to OPP and can cause the same effects.

OSHA See Occupational Safety and Health Administration.

Oxygen deficient Oxygen level below 19.5 percent.

Oxygen enriched Oxygen level above 23.5 percent.

Permissible exposure limit (PEL) OSHA value that regulates the amount of chemical that a person can be exposed to during an 8-hour day.

pH An abbreviation for potential of hydrogen, power of hydrogen, percentage of hydrogen ions, and the negative logarithm of hydrogen ion activity.

Photoionization detector (PID) A detector that measures organic materials and some inorganics in air by ionizing the gas with an ultraviolet lamp.

pH paper Testing paper used to indicate the corrosiveness of a liquid.

Polymerase chain reaction (PCR) Biological detection device that examines the DNA of the suspect material.

Polytech The name for Gastec's unknown gas detector tube.

Polytest The name for the Dräger unknown gas detector tube.

Precision The ability of a detector to repeat the results for a known atmosphere.

Prime Alert Protein screen for biological materials.

Protein screen Test for material that contains protein, which could be a biological agent.

Quality factor Factor used to determine potential biological damage from radiation and to convert values from absorbed dose to equivalent dose.

Radiation absorbed dose (RAD) A quantity of radiation.

Radiation pager Detection device that signals in the presence of gamma and X-rays.

Radioactive decay Emission of radioactive energy.

Radioisotope Isotope whose nucleus is unstable and that emits radioactivity to regain stability.

Radiological dispersion device (RDD) Explosive device that spreads a radioactive material. The explosive power is derived from a non-nuclear source, such as a pipe bomb, which is attached to a radioactive substance.

Radon Common radioactive gas found in homes.

Raid-M-100 Warfare agent detection device.

Raman spectroscopy Detection technology in which a laser is used to excite the sample.

RAMP Biological testing device.

RBR See Risk-based response.

Reagent A chemical material (solid or liquid) that is changed (reacts) when exposed to a chemical substance.

Recommended exposure level (REL) Exposure value established by NIOSH for a 10-hour day, 40-hour work- week. Is similar to PEL and TLV.

Recovery time The amount of time it takes for a detector to return to 0 after exposure to a gas.

Relative gas density (RgasD) Term used by the NIOSH Pocket Guide that means vapor density. It is a comparison of the weight of a gas to the weight of air.

Relative response factor See Correction factor.

Risk-based response A response methodology that characterizes risk.

Risk-based response A system for identifying the risk chemicals present even though their specific identity is unknown. This system categorizes all chemicals into fire, corrosive, or toxic risks.

Roentgen Basic unit of measurement for radiation.

Roentgen equivalent man (REM) Method of measuring radiation dose.

SABRE 5000 Warfare agent detection device that detects explosives, drugs, chemical warfare agents, and toxic industrial materials.

Safety Data Sheet (SDS) A form with chemical information that lists properties, health effects, and emergency actions.

Scintillation Radiation detection technology.

Sensitizer Chemical that after repeated exposure may cause an allergic effect for some people.

Short-term exposure limit (STEL) Fifteen-minute exposure to a chemical; requires a 1-hour break between exposures and is only allowed four times a day.

Sick building A building that is suspected of causing irritation or health problems for the occupants.

Sick building syndrome (SBS) A temporary condition affecting some building occupants with symptoms of acute discomfort that stop when the person leaves the building.

Sievert (Sv) Unit of measurement equivalent to REM.

SLUDGEM Acronym for salivation (drooling), lacrimation (tearing), urinary, defecation, gastrointestinal upset, emesis (vomiting), and miosis (pinpointed pupils).

Smart sensor A sensor that has a computer chip on it that allows the switching of a variety of sensors within an instrument.

Sniffex Explosives detection device.

Solid polymer electrolyte (SPE) A sensor that uses a membrane to diffuse oxygen through water to measure the oxygen content. Also known as an oxygen sensor.

Surface acoustic wave (SAW) A sensor technology used in detection devices, used in the SAW and hazardous materials CADS. After the sampled gas passes over the sensor, it outputs an algorithm that is checked for a possible match.

Technical Hazards Response Unit (THRU) Specialized response unit within the FBI for WMD and other high-hazard crime scenes.

Thermal conductivity sensor A type of LEL sensor.

Thermoluminescent dosimeter (TLD) Radiation monitoring device worn by responders.

Threshold limit value (TLV) Exposure value that is similar to the PEL but issued by the ACGIH. It is for an 8-hour day.

Toxic gas sensor Device for the detection of toxic gases. Common toxic gas sensors are for CO, H_2S, ammonia, and chlorine.

Toxicology The study of poisons.

Toxic twins Carbon monoxide and hydrogen cyanide, which are commonly found in fire smoke and have a synergistic effect on the body.

TRACEM Acronym for the types of hazards that may exist at a chemical incident; thermal, radiation, asphyxiation, chemical, etiological, and mechanical.

UEL See Upper explosive limit.

Upper explosive limit (UEL) The upper limit of the flammable range; the maximum concentration of flammable gas or vapor that can be mixed with air to have a fire or explosion.

Vapor density (VD) The weight of a gas compared with an equal amount of air. Air is given a value of 1. Gases with a vapor density less than 1 rise, while those with a VD greater than 1 sink.

Vapor pressure The force of vapors coming from a liquid at a given temperature.

Volatility The amount of vapors coming from a liquid.

VX A nerve agent, more toxic than sarin.

Wastewater strip A strip that does some additional tests beyond the chemical classifier.

Wheatstone bridge A form of LEL sensor.

X-ray A form of radiation much like light and gamma radiation but that bombards a target with electrons, which makes it very penetrating.

Index

Note: Page numbers followed by *f* indicate figures; those followed by *t* indicate tables.